PRINCIPLES AND APPLICATIONS OF PHOTOCHEMISTRY

Principles and Applications
of
Photochemistry

RICHARD P. WAYNE

Dr Lee's Reader in Chemistry, Christ Church, Oxford

and

University Lecturer in Physical Chemistry, University of Oxford

Oxford   New York   Tokyo

OXFORD UNIVERSITY PRESS

1988

o3317948

CHEMISTRY

Oxford University Press, Walton Street, Oxford OX2 6DP
Oxford   New York   Toronto
Delhi   Bombay   Calcutta   Madras   Karachi
Petaling   Jaya   Singapore   Hong Kong   Tokyo
Nairobi   Dar es Salaam   Cape Town
Melbourne   Auckland
and associated companies in
Berlin   Ibadan

Oxford is a trade mark of Oxford University Press

Published in the United States
by Oxford University Press, New York

© Richard P. Wayne, 1988

An earlier version of this work was
published as Photochemistry, 1970, by
Butterworths, London
This edition 1988

British Library Cataloguing in Publication Data

Wayne, Richard P. (Richard Peer)
Principles and applications of photochemistry
1. Photochemistry
I. Title   II. Wayne, Richard P. (Richard Peer). Photochemistry
541.3'5
ISBN 0–19–855234–3
ISBN 0–19–855233–5 Pbk

Library of Congress Cataloging in Publication Data

Wayne, Richard P. (Richard Peer)
Principles and applications of photochemistry.
Rev. ed. of: Photochemistry. 1970.
Bibliography: p. 268 Includes index.
1. Photochemistry. I. Wayne, Richard P. (Richard Peer). Photochemistry. II. Title.
QD708.2.W39   1988   541.3'5   88–5283
ISBN 0–19–855234–3
ISBN 0–19–855233–5 (pbk.)

Set by Macmillan India Ltd, Bangalore-25

Printed and bound in Great Britain by
Biddles Ltd, Guildford and King's Lynn

PREFACE

Photochemical processes are of the greatest importance to life on Earth. Photosynthesis harvests the Sun's energy and creates carbohydrate from atmospheric carbon dioxide, as well as liberating oxygen to the atmosphere. Light-induced chemical changes in the gases of the atmosphere and the particles suspended there also modify the chemical composition of the atmosphere and allow it to support life. Indeed, the formation from the simplest elements of the complex organic molecular precursors of life, and then the emergence of life itself, are intimately bound up with photochemical processes. One of the most important senses for Man, and many other species besides, is vision, which is also photochemical in origin. Nature thus relies on light to effect some of her most essential chemistry. Man has also tried to harness light in his service, the applications ranging from the synthesis of new and complex organic species, through numerous kinds of imaging and photographic process, to the gathering and storage of solar energy.

The significance of photochemistry is by no means limited to the use that Nature and Man make of it. Rather, the chemistry itself is of profound interest at the most fundamental level. The reactions, dissociations, isomerizations, and optical emissions of electronically excited species are the central feature of photochemistry. For every atom and molecule known to us in the ground state, there are likely to be one or more excited states. Since these states possess different electronic structures from, and higher energies than, their parents, their chemistry is almost inevitably distinct from that of the ground-state species. Whole new fields of chemistry are thus opened up by allowing light to interact with the elements and their compounds. The motivations for studying photochemistry are as diverse as those for studying chemistry itself. On the one hand, physical chemists may be interested in the detailed dynamics of a photodissociation process and of the progress of the changes on a time scale of less than a picosecond, while on the other, organic chemists may seek an improved understanding, through the examples that photochemistry affords, of the relationships between reactivity and electronic and molecular structure.

In 1970, I wrote a textbook on photochemistry, with the undergraduate or non-specialist research student in mind (*Photochemistry*, Butterworths). At that time, the only available textbooks were twenty years out of date, with the exception of the monumental reference work *Photochemistry* by J. G. Calvert and J. N. Pitts, Jr (John Wiley, New York, 1966). My book was written from the standpoint of physical chemistry, although there was much of organic and inorganic chemistry within. Since 1970, a steady stream of other excellent books on photochemistry has appeared, but none with quite the same emphasis that I had adopted, and it seemed that the time was now ripe to write a new book along the same lines as my earlier one. The present volume is the result. It is intentionally shorter than the 1970 *Photochemistry*, containing only the most essential material. At the same time as shortening most parts of the book, I have expanded the final chapter on the applications of photochemistry. That chapter was always popular with readers, and many new developments have arisen since 1970. I believe that it is worth weighting the balance of the book towards these practical examples, because they illustrate admirably the fundamental concepts of photochemistry, and because most chemists will ultimately be most involved with photochemistry in the applied context.

Photochemistry has seen an enormous increase of research activity in all areas over the past twenty years. There are now three Photochemical Societies, in Europe, America, and Japan. At least two journals are devoted to photochemistry (the *Journal of Photochemistry* [now *Journal of Photochemistry and Photobiology*], which started in 1970, and *Photochemistry and Photobiology*). There are regular international and informal conferences on photochemistry, together with further photochemical conferences held under the auspices of the International Union of Pure and Applied Chemistry (IUPAC). A series of short reports on the state of knowledge in various parts of photochemistry, written by acknowledged leaders, was edited by Dr John Coyle and myself as *Photochemistry: Past, Present and Future*, a special issue of *Journal of Photochemistry* in celebration of the publication of Volume 25 in 1984.

The huge research effort has led to an explosion of knowledge that has made the task of selecting material for a book even harder than it was in 1970, and the problem is compounded by the numerous new and important techniques used for the study of various aspects of photochemistry that need a mention if the reader is to grasp the nature of modern photochemistry. The increased knowledge makes difficult not so much deciding which principles must be expounded, but rather the choice of examples that will illustrate them. Very often, the simplest examples, and those that are understood in the greatest detail, involve atoms or small molecules in the gas phase. A physical chemist naturally finds such examples appealing, but he must be aware that most chemists work with large molecules in condensed-phase systems. I hope that I have selected my examples in a balanced and sensible way!

The structure of this book is very similar to that of the 1970 *Photochemistry*. Photochemistry is concerned with the chemistry of excited species, and the subject is developed to show the several paths by which an excited species may react or undergo radiative or radiationless decay. Chapters 3–6 describe these paths; Chapter 1 gives a general introduction to the basic concepts of photochemistry, while Chapter 2 explains briefly the principles of absorption and emission of radiation. Fairly obviously, there are certain experimental techniques peculiar to photochemistry, and the descriptive material is placed in a better context if the reader understands how the experiments were performed. I have explained some of the more important techniques in Chapter 7. I have left this discussion until quite late on in the book, because the theoretical background needs to be developed first. Chapter 7 serves also as a vehicle for a very brief presentation of what has been called 'High-resolution Photochemistry'. This field is concerned with the detailed dynamics of photochemical processes, including the utilization of energy in specific quantum states of the starting species and its disposal in the products. It is proving very successful in probing the intimate nature of photochemical interactions, but does not fit easily into the main development of Chapters 3–8. Since the experimental part of the work is very technique oriented, it seems appropriate to describe the methods in Chapter 7, and to alert the reader here to the existence of the material in Section 7.6, and the later parts of Section 7.5. Chapter 8 concludes the book with the discussion, mentioned earlier, of photochemical processes found in nature and of some commercial and laboratory applications. Incidentally, in this last chapter I have made no effort to stick rigidly to systematic names for chemical compounds, but rather thought it sensible to use the names almost universally employed in industry.

This book is intended primarily for undergraduate readers, although it is hoped that it may prove useful and interesting to graduate students embarking on research in photochemistry. Some knowledge of elementary chemistry (e.g. atomic and molecular structure, spectroscopy, reaction kinetics) is assumed, but the ideas of photochemistry are built up from first principles. Specific literature references are out of place in a book of this kind. Instead, a bibliography is given at the end of each chapter to enable the reader to pursue in greater depth the topics discussed. These bibliographies have deliberately been made more extensive for the later chapters in the book, especially for Chapters 7 and 8. Articles relevant to the overall aims of each chapter are followed by articles on specific topics identified, where possible, by their section numbers. Series of review volumes are an important additional source of information. They include the Specialist Periodical Reports in Photochemistry (Royal Society of Chemistry: Senior reporter D. Bryce Smith), *Advances in Photochemistry* (present editors: D. H. Volman, K. Gollnick, and G. S. Hammond), which is biased towards physical chemistry, and *Organic Photochemistry* (present editor: A. Padwa), whose title defines its interests. Excited states are treated specifically in the series *Excited States* (edited by E. C. Lim) from a theoretical and physical point of view, and a whole volume of *Advances in Chemical Physics* (Vol. 50, 1982, edited by K. P. Lawley) was devoted to the dynamics of the excited state. One volume that is particularly useful for supplementary reading in connection with Chapter 8 is by J. D. Coyle, R. R. Hill, and D. R. Roberts (eds), *Light, Chemical Change and Life*, The Open University Press, Milton Keynes (1982). This book, which was written as a source text for a third-level Open University course in photo-chemistry, contains fairly short, readable, and authoritative accounts of a wide variety of topics concerning the biological and technological applic-ations of photochemistry.

I recorded, in 1970, my debt of gratitude to Professor R. G. W. Norrish, FRS, who was my first teacher, and to Dr E. J. Bowen. Both were outstanding pioneers in photochemistry, representing the Cambridge and Oxford schools. Sadly, both have died since 1970. Others whose help I acknowledged in-cluded Professors C. H. Bamford, FRS, J. N Pitts, Jr, and B. A. Thrush, FRS, and I should like to reiterate here my appreciation of the help they have always given to me. To their names I should like to add those of Professors W. J. Albery, FRS and R. J. Donovan, Drs N. S. Allen and J. D. Coyle (who are my Associate Editors on the *Journal of Photochemistry and Photobiology*, and who gave me particularly useful advice about Chapter 8), and my Oxford photochemical colleagues, in particular Drs Gus Hancock and Mike Pilling.

Finally, I should like to thank my wife, Brenda, not only for her tolerance while I was writing the book, but also for giving me the benefit of her invaluable editorial skills.

Oxford   R. P. W.
October 1987

ACKNOWLEDGEMENT

Figure 8.13 is from Eaton, D. F. (1986). Dye sensitized polymerization. In *Advances in photochemistry*, volume 13, (ed. D. H. Volman, G. S. Hammond, and K. Gollnick), pp. 427–487, Copyright © 1986 by John Wiley & Sons, inc., reprinted by permission of John Wiley & Sons, inc.

CONTENTS

Contents   xiii

1

Basic principles of photochemistry

1.1 Scope of photochemistry

Man has been aware from the earliest times of the influence that the Sun's radiation has on matter. However, it is during the last seventy years or so that a systematic understanding of photochemical processes has developed. A logical pattern to the interaction between light and matter emerged only after the concept of the quantization of energy was established. It is the purpose of this book to explain the physical foundations on which modern photochemistry is based; the specific examples given in the book are intended to illustrate these principles rather than to provide a comprehensive survey of known photochemical reactions.

'Photochemistry' is a term rather loosely applied. While an important part of photochemistry is concerned with the chemical change that may be brought about by the absorption of light, a number of physical processes that do not involve any overall chemical change lie within the province of the photochemist; processes such as *fluorescence* (in which light is emitted from a species that has absorbed radiation) or *chemiluminescence* (in which light is emitted as a 'product' of a chemical reaction) must be regarded as of a photochemical nature. The word 'light' is also used loosely, since radiation over a far wider range of wavelengths than the visible spectrum is involved in processes that would be accepted as photochemical. The long wavelength limit is probably in the near infrared (say at 2000 nm) and the region of interest extends into the vacuum ultraviolet (see footnote p. 155) and is limited only formally at the wavelengths where radiation becomes appreciably 'penetrating' (X-rays). The essential feature of photochemistry is probably the way in which 'excited' states of atoms or molecules play a part in the processes of interest. It is apparent that absorption or emission of radiation to or from these states is the concern of the spectroscopist as well as of the photochemist, and the photochemist must have at least a background knowledge of spectroscopy. At the same time, the photochemist is frequently interested in the *rates* at which processes occur, so that the concepts of *reaction kinetics* are often employed. It is assumed that the reader of this book has had contact with the ideas of quantum theory, spectroscopy, and reaction kinetics, and that he can obtain access elsewhere to more detailed discussions of these topics than it is possible to provide here.

1.2 Light and energy

Planck developed his theory of black-body radiation on the basis of a postulate that radiation possessed particulate properties and that the particles, or *photons*, of radiation of specific frequency v had associated with them a fixed energy ε given by the relation

$$\varepsilon = hv \qquad (1.1)$$

where h is called Planck's constant. This quantum theory of radiation was then used by Einstein to interpret the photoelectric effect. As early as the beginning of the nineteenth century, Grotthus and Draper had formulated a law of photochemistry which stated that only the light absorbed by a molecule could produce photochemical change in the molecule. The development of the quantum theory led to a realization that the radiation would be absorbed in the quantized energy packets; Stark and Einstein suggested that one, and only one, photon was absorbed by a single particle to cause its photochemical reaction. It is now appreciated that several processes may compete with chemical reaction to be the fate of the species excited by absorption (see Section 1.5) and a more satisfactory version of the Stark–Einstein law states that *if a species absorbs radiation, then one particle is excited for each quantum of radiation absorbed.* Although this law might appear trivial in the present-day climate of acceptance of the quantum theory, the law is of fundamental importance in photochemistry, and the agreement between experiment and predictions based on the law does, in fact, offer substantial evidence in favour of the quantum theory of radiation.[†]

It is now apparent that the energy of excitation of each absorbing particle is the same as the energy of the quantum given by the Planck relation (1.1), and the excitation energy per mole is obtained by multiplying this molecular excitation energy by N, Avogadro's number. A linear relationship exists between energy and frequency, so that frequency characterizes radiation in a particularly direct way. It has been, however, almost universal practice to discuss the visible and ultraviolet regions of the spectrum in terms of *wavelength* of the radiation, and it is therefore convenient to express the molar

[†] A number of photochemical processes are recognized in which more than one quantum of radiation is absorbed by a single molecule. Many such processes do not violate the Stark–Einstein law: they involve excitation to successively higher energy states of the molecule, each step requiring a single quantum. True 'simultaneous' absorption of more than one photon occurs under conditions of intense illumination, as described in Section 3.9. Observation of such multiphoton processes awaited the development of suitably intense sources, especially lasers. Light is still absorbed in quantized packets.

excitation energy, E, in terms of wavelength, λ,

$$E = Nh\nu = \frac{Nhc}{\lambda} \qquad (1.2)$$

where c is the velocity of light. Numerical relationships between E and λ may be derived from the values given for constants in Appendix 1; one particularly useful form is

$$E = \frac{119\,627}{\lambda} \text{ kJ mol}^{-1} \qquad (1.3)$$

where λ is in nanometres.

Although the chemist frequently finds thermal energy units (kJ) most useful, it is sometimes convenient to express energies in units such as electron-volts or wavenumber: the conversion factors are given in Appendix 2. The relationships are displayed pictorially inside the front cover. A useful way of remembering the approximate energies of photochemically active radiation is to recall that the wavelength range is roughly 200–600 nm, while the corresponding energies are in the range 600–200 kJ mol^{-1}.

1.3 Excitation by absorption

A molecule that has absorbed a quantum of radiation becomes 'energy-rich' or 'excited' in the absorption process. Absorption in the wavelength region of photochemical interest leads to *electronic excitation* of the absorber. Absorption at longer wavelengths usually leads to the excitation of vibrations or rotations of a molecule in its ground electronic state. Although it would be a mistake to suppose that the *only* form of excitation that could lead to photochemical change is electronic, it is generally true that electronically excited states are involved in photochemical processes. The importance of electronic excitation is, in part, a result of the energy possessed by the 'hot' species, as we shall see in the next section. There is, however, another reason, connected with the nature of the excitation, why electronically excited species exhibit reactivities distinguishable from those of the unexcited species. A simple example will make this clear. The electronic structure of lithium in its ground state is represented in the familiar form Li $1s^2\,2s^1$: the electrons are placed in the lowest orbitals available to them. The configuration Li $1s^2\,3p^1$ lies at a higher energy than the $1s^2\,2s^1$ configuration, and represents an electronically excited state of lithium. Now, this excited lithium atom possesses an electron in a p orbital. Since the chemistry of a species depends to a considerable extent on the electronic structure, the reactivity of the excited lithium atom can be expected to differ from that of the ground-state atom, quite apart from considerations of the extra energy possessed in the excited configuration. Rather more subtle forms of electronic excitation are possible, and these will be explored further in Chapter 2. For any one atomic or molecular species, a great many electronically excited states may be accessible: for each of these states there may be a distinct chemistry, which is by no means identical with that of the ground state. Thus, it is apparent that the reactions observed in photochemical studies may have little in common with the thermal reactions of the parent, ground-state species.

1.4 Thermal and photochemical reactions

The essential distinction between thermal and photochemical reactions now needs to be explored more fully. Thermal energy may be distributed about all the modes of excitation in a species: in a molecule these modes will include translational, rotational, and vibrational excitation, as well as electronic excitation. However, for species in thermal equilibrium with their surroundings, the Boltzmann distribution law is obeyed. This law states, of course, that the relative numbers of particles, n_1 and n_2, in two equally degenerate levels 1 and 2, separated by an energy gap ΔE, are given by the expression

$$\frac{n_2}{n_1} = e^{-\Delta E/RT} \tag{1.4}$$

If we take a typical energy of an electronically excited state equivalent in thermal units to $250 \, \text{kJ} \, \text{mol}^{-1}$, we can show that at room temperature $(RT \sim 2500 \, \text{J} \, \text{mol}^{-1})$ $n_2/n_1 = e^{-100} \sim 4 \times 10^{-46}$, so that a negligible fraction of the species is excited. To achieve a concentration of only 1% of the excited species would require a temperature of around $6800 \degree C$, and in the event most *molecular* species would undergo rapid thermal decomposition from the ground electronic state and it would not be possible to produce appreciable concentrations of electronically excited molecules. In contrast, if molecules absorb radiation at a wavelength of about 500 nm as a result of an electronic transition, then electronic excitation certainly *must* occur, and the concentration produced depends on several factors, including the intensity of illumination and the rate of loss of the excited species. That chemical change can follow the production of electronically excited species is not surprising when consideration is given to the energies involved. The very rough wavelength range suggested in Section 1.1 as being of photochemical interest is similar in equivalent energy to the values of chemical bond energies normally encountered. If the electronic excitation energy can in some way be made available for bond rupture, then chemical change may occur. The study of photochemistry is concerned in part with the manner and extent of such energy utilization. Again, the energy of excitation is comparable with the activation energies frequently observed for the reaction of unexcited species: if the electronic excitation can be used to overcome all or some of the energy of activation, then it may be expected that the excited species will react *more rapidly* than the ground-state species. Thus we see that photochemical reactions are distinguished from thermal reactions, first by the relatively large concentrations of highly excited species, which may react faster than the ground-state species and may even participate isothermally in processes that are endothermic for the latter, and secondly, if the excitation is electronic, by the changes in chemical reactivity that may accompany the new electronic configuration of the species.

A secondary feature of photochemical excitation is that a specific state of the species is formed if the radiation is contained within a sufficiently narrow band: an essentially monoenergetic product can result. It is true that the species may still possess, about its excited level, an energy spread character-istic of the temperature of the surroundings, but at room temperatures the range of energies within which most particles lie is very small compared with the energy of excitation, and even narrower distributions may be achieved at reduced temperatures. The possibility of forming monoenergetic species is of particular concern in connection with theories of reaction kinetics, where it is of fundamental interest to see how rapidly a species possessing a specific amount of energy can participate in a reaction. Monoenergetic species can be produced thermally only by the use of sophisticated methods such as the molecular beam technique, while simple photochemical experiments can achieve a relatively narrow energy distribution for electronically excited species.

1.5   Fates of electronic excitation

Photochemical processes involving the absorption of light can be divided into the act of absorption, which falls within the domain of spectroscopy, and the subsequent fate of the electronically excited species formed. It has been implied, in the discussion of the last few pages, that there are several such fates, and we shall now consider more explicitly what the possibilities are. At this stage a highly simplified picture will be presented: each of the processes mentioned will be explored in greater detail later in the book.

Figure 1.1 represents, in simplified form, the various paths by which an electronically excited species may lose its energy.

Energy transfer, represented by paths (iv) and (v) in the diagram, leads to excited species, which can then participate in any of the general processes. In this preliminary discussion, therefore, energy transfer will not be considered further; instead, this topic is deferred until Chapter 5.

Chemical change can come about either as a result of dissociation of the absorbing molecule into reactive fragments (process i), or as a result of direct reaction of the electronically excited species (process ii); electronically excited

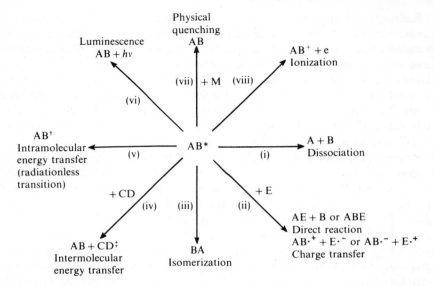

Fig. 1.1 The several routes to loss of electronic excitation. The use of the symbols *, †, and ‡ is only intended to illustrate the presence of electronic excitation and not necessarily differences in states. One or both of the products of processes (i)–(iii) may be excited.

species may also undergo spontaneous isomerization, as indicated by path (iii). Some examples will illustrate these processes; an asterisk is used to denote electronic excitation.

$$NO_2^* \rightarrow NO + O \quad [\text{Dissociation}] \tag{1.5}$$

$$O_2^* + O_3 \rightarrow 2O_2 + O \quad [\text{Reaction}] \tag{1.6}$$

cis-$C_6H_5CH{=}CHC_6H_5^*$
stilbene

trans-$C_6H_5CH{=}CHC H_5$
(Isomerization)

dihydrophenanthrene

$$(1.7)$$

Several mechanisms for dissociation are recognized (they include *optical dissociation, predissociation, induced predissociation*, and so on), and they are discussed in more detail in Chapter 3. A special case of dissociation is that of ionization, shown as path (viii).

Radiative loss of excitation energy (path vi) gives rise to the phenomenon of *luminescence*: the terms *fluorescence* or *phosphorescence* are used to describe particular aspects of the general phenomenon. Luminescence is subject to the laws of radiative processes, and it is more conveniently treated after the discussion of Chapter 2.

Path (vii) indicated in Fig. 1.1 is *physical quenching*. In this process an atom or molecule M can relieve AB* of its excess energy. Physical quenching differs only formally from intermolecular energy transfer in that M, which must initially take up some excitation energy, does not make its increased energy felt in terms of its chemical behaviour. The electronic excitation of AB* is, in fact, frequently converted to translational or vibrational excitation of M.

1.6 Secondary reactions: intermediates

We have considered, so far, only the immediate fate of electronic excitation in an absorbing species. It is obvious that the products of processes (i)–(iii) (for example, the oxygen atoms produced in reaction 1.5 or 1.6) can themselves participate in chemical reactions which are, therefore, a direct consequence of the initial absorption of a quantum of radiation. It is usual to differentiate between these *secondary* reactions and the *primary* photophysical processes. As an illustration of these steps, consider the photolysis of ketene (*photolysis* is used to describe a process in which the absorption of light is followed by the rupture of a bond):

$$CH_2CO + h\nu \rightarrow CH_2CO^* \quad \left.\vphantom{\begin{array}{c} 1 \\ 1 \end{array}}\right\} \quad \text{Primary process} \tag{1.8a}$$

$$CH_2CO^* \rightarrow CH_2 + CO \tag{1.8b}$$

$$CH_2 + CH_2CO \rightarrow C_2H_4 + CO \quad \text{Secondary process} \tag{1.9}$$

Methylene (CH_2) formed in the primary process subsequently reacts with ketene to yield ethene and carbon monoxide.

Secondary reactions are the ordinary thermal reactions of the various participating species; they are photochemical only in the sense that the reactive species would not have appeared in the absence of light. Certain types of chemical species are found as intermediates more commonly in photochemical than in thermal reaction systems, largely as a result of the energies involved. These intermediates include free atoms and radicals, as well as electronically excited species. Such intermediates are frequently highly reactive, and their lifetime in a reaction system is correspondingly short. However, reactivity must not be confused with instability: a free radical or atom *in isolation* would have perfectly normal stability, while the lifetime of an electronically excited species would be determined by the probability of losing energy by radiation. Atoms and radicals may themselves carry some kind of excess energy: for example, photolysis of ketene may yield ground-state or electronically excited methylene, depending on the wavelength of radiation absorbed.

Chain reactions are very typical of atoms and radicals, and such processes are frequently encountered in photochemistry. An example of a rapid chain reaction is the photochemically initiated hydrogen–chlorine explosion:

$$Cl_2 + h\nu \rightarrow Cl + Cl \tag{1.10}$$

$$Cl + H_2 \rightarrow HCl + H \tag{1.11}$$

$$H + Cl_2 \rightarrow HCl + Cl \tag{1.12}$$

Here, reaction (1.10) describes the primary process up to the formation of chlorine atoms; the chain propagation steps are secondary reactions, and both atomic hydrogen and chlorine are *chain carriers*.

1.7 Quantum yields

A concept of great value in photochemistry is that of the *quantum yield, ϕ*. As originally understood, it was *the number of molecules of reactant consumed for each photon of light absorbed*. In this form the quantum yield reflects, without distinction, both the efficiency of the primary photochemical process in bringing about chemical change and also the extent of secondary reaction. A quantum yield greater than unity suggests the occurrence of secondary reactions, since the Stark–Einstein law indicates that not more than one molecule can be decomposed in the primary step (a quantum yield greater than two points to the operation of a chain reaction mechanism). However, the discussion of Section 1.5 will have shown that chemical change is not the only consequence of absorption of radiation. Thus, a chain reaction may be taking place in a photochemical reaction even though the overall quantum yield is less than unity. It is more helpful to consider primary and overall quantum yields, ϕ and Φ, separately; the primary quantum yield should be stated for a specific primary process (in relation to Fig. 1.1, to one of the paths i-viii). If, for example, dissociation precedes secondary chemical reactions, the primary quantum yield would be the number of molecules dissociating in the primary step for each quantum of light absorbed, and the ratio of overall to primary quantum yields then indicates the extent of secondary reaction. Where nothing to the contrary is stated, overall quantum yield refers to the removal of reactant, although, if several different secondary paths exist, it may be desirable to quote an overall quantum yield for the formation of a specific product.

The determination of overall quantum yields for chemical change requires measurement of the numbers of molecules of reactant consumed, or of product formed, and of the number of quanta of radiation absorbed. The former measurement just involves suitable analytical techniques, while the latter requires a method for measuring absolute numbers of photons. The experimental procedures adopted for such absolute measurements are described in Chapter 7. In the determination of primary quantum yields, the contribution to chemical change of secondary reactions must first be eliminated or allowed for, and the absolute efficiencies of radiative and non-radiative energy-loss processes must be assessed. It is not always possible even to establish what primary paths exist, so that a full description of the primary processes in terms of quantum yields can be made only in favourable cases. Nevertheless, several indications may be used to gain some insight into the primary process. The nature of the absorption spectrum may suggest the electronic configuration of the excited state and, hence, the probable fate of the energy. Detection of the intermediates (excited states as well as atoms and radicals) may reveal the products of the primary step (see, for example, Section 7.4). Measurement of overall quantum yields can also give some information about the primary process. If $\Phi \ll 1$, then in all probability little chemical change occurs in the excited absorbing molecule (although 'cage' recombination of radicals in condensed-phase reaction systems is another very common cause of low quantum yields; cf. Section 3.7). A search must then be made for radiation emitted from the system; the spectrum will indicate whether the emission is fluorescence of the absorber or whether it is derived from a state populated by intermolecular or intramolecular energy transfer. Study of *fluorescence quenching* (Chapter 4) will yield information about physical deactivation processes. Again, if the quantum yield for formation of a specific product is invariant with experimental conditions, such as reactant concentrations or temperature, then that product probably appears, at the measured efficiency, in the primary process.

The energy of an excited species must go somewhere, so the Stark–Einstein law leads to the conclusion that the sum of the quantum yields for *all* primary processes, including deactivation, must be unity. Where sufficient experimental data are available, this expectation is well substantiated.

1.8 Reaction kinetics

The ratio of overall to primary quantum yields, Φ/ϕ, is analogous to the kinetic chain length, v, determined in studies of thermal chain reactions. The quantities may be expressed in terms of rate constants for the several secondary reactions, and their variation with concentrations of various species may lead to confirmation of a hypothetical reaction mechanism and evaluation of rate constants.

In thermal reactions, v is defined by the relation

$$v = \frac{\text{rate of reactant disappearance}}{\text{rate of initiation}} \qquad (1.13)$$

Quantum yields may also be defined in terms of *rates*, rather than *numbers* of molecules and photons. An *intensity* of radiation, I, refers to an energy per unit time, and it is frequently convenient to express the *absorbed intensity*, I_{abs}, as the energy absorbed in unit time by unit volume: it is then in the same form as a rate expressed in concentration units, with energies (numbers of photons) replacing numbers of molecules. Hence, for a process

$$A + hv \rightarrow \text{products} \qquad (1.14)$$

$$\Phi = \frac{-d[A]/dt}{I_{abs}} \qquad (1.15)$$

Furthermore, if we assume that the primary quantum yield, ϕ, is for formation of reactive intermediates, then ϕI_{abs} is the rate of initiation in the photochemical system, and

$$\frac{\Phi}{\phi} = \frac{\text{rate of reactant disappearance}}{\text{rate of initiation}} \equiv v \qquad (1.16)$$

In fact, initiation by photochemical means is often the best way in which to study the kinetics of radicals or energy-rich species, since not only may the rate of initiation be measured accurately, but also the temperature at which the experiment is performed may be sufficiently low to prevent the occurrence of a plethora of confusing processes often found in thermal reactions initiated at high temperatures. A very simple example will show the use of quantum yield measurements in the elucidation of reaction mechanisms and rate constants. The photolysis of ozone–oxygen mixtures by red light might be expected to proceed via the mechanism

$$O_3 + hv \xrightarrow{\phi_1} O_2 + O \qquad (1.17)$$

$$O + O_3 \xrightarrow{k_2} 2O_2 \qquad (1.18)$$

$$O + O_2 + M \xrightarrow{k_3} O_3 + M \qquad (1.19)$$

The 'third body', M, is necessary in reaction (1.19), as in many other atom recombination reactions, to stabilize the vibrationally 'hot' molecule formed immediately after recombination.

The rate equations for the formation of O and loss of O_3 are

$$\frac{d[O]}{dt} = \phi_1 I_{abs} - k_2[O][O_3] - k_3[O][O_2][M] \qquad (1.20)$$

$$-\frac{d[O_3]}{dt} = \phi_1 I_{abs} + k_2[O][O_3] - k_3[O][O_2][M] \qquad (1.21)$$

Multistep reaction schemes are interpreted kinetically by writing down the differential equations, such as eqn (1.20) or eqn (1.21), for all the species of interest, including the intermediates. Solution of these equations then allows prediction of the concentration–time variation of each of the species. Unfortunately, analytical solution of the many simultaneous differential equations is rarely possible. Numerical solution has become a widely used alternative since the advent of high-speed computers and of good techniques for dealing with differential equations. Such methods do not, however, afford much insight into the underlying chemistry of the system. For some highly reactive intermediates, the *Stationary State Hypothesis* (SSH) provides a simplification that will permit algebraic solutions of the kinetic equations. Consider an intermediate X that is created in a process whose rate is constant, and whose loss rate increases with increased [X]. After the reaction is started, [X] will increase until the rate of loss is equal to the rate of formation: a steady state for [X] has been reached, and $d[X]/dt$ is very nearly zero.

In applying the SSH to our example of ozone photolysis, we set the differential in eqn (1.20) to zero, since atomic oxygen is a highly reactive intermediate. To simplify the appearance of the equations, let us rewrite eqn (1.20) in the form

$$\frac{d[O]}{dt} = \phi_1 I_{abs} - k'[O] \tag{1.22}$$

where $k' = k_2[O_3] + k_3[O_2][M]$. Thus, setting this equation to zero yields the value for the steady-state concentration of oxygen

$$[O]_{ss} = \phi_1 I_{abs}/k' \tag{1.23}$$

This value for the atomic oxygen concentration might then be used for substitution in eqn (1.21) to obtain the rate of ozone loss. The problem is to know whether the concentration of O calculated using the SSH bears any relation to actual concentrations. Our example has been chosen because it can also be solved analytically. So long as the extent of photolysis is small, I_{abs}, $[O_3]$, $[O_2]$, and $[M]$ can all be taken to be nearly independent of time, and eqn (1.22) can be integrated to yield

$$[O] = \frac{\phi_1 I_{abs}}{k'}[1 - \exp(-k't)] \tag{1.24}$$

where t is the time for which the system has been illuminated. This equation approaches the steady-state expression if $k't \gg 1$, the error in applying the SSH being $< 1\%$ for $k't \geqslant 4.6$. In a typical study, the concentration of $[O_3]$ might be 5×10^{-5} mol dm^{-3}, and that of $[O_2]$ might be 10^{-3} mol dm^{-3}. The measured values of k_2 and k_3 are about 4×10^6 dm^3 mol^{-1}s^{-1} and 2×10^8 dm^6 mol^{-2}s^{-1}, respectively, for a temperature of 18°C. Thus, the value of k' would be approximately 450 s^{-1}, and the steady-state concentration of [O] will be approached in something like $4.6/400 \simeq 0.01 \text{ s}$. This conclusion suggests that the steady-state treatment would be perfectly adequate to deal with an experiment in which ozone was being photolysed slowly over a period of minutes or hours. We should recognize, however, that the treatment would not be satisfactory if the concentrations of $[O_3]$ and $[O_2]$ were several orders of magnitude smaller than those postulated, or if it were desired to make kinetic measurements on a smaller time scale. In principle, the applicability of the SSH should be established for any particular kinetic problem before it is used.

In our example, we *have* established that the steady-state treatment will be adequate, so that eqn (1.23) may be substituted in eqn (1.21) to yield the relation

$$-\frac{d[O_3]}{dt} = \frac{2\phi_1 I_{abs} k_2 [O_3]}{k_2[O_3] + k_3[O_2][M]} \tag{1.25}$$

or

$$\Phi = -\frac{d[O_3]/dt}{I_{abs}} = \frac{2\phi_1 k_2 [O_3]}{k_2[O_3] + k_3[O_2][M]} \tag{1.26}$$

Inversion of eqn (1.26) yields the expression

$$\frac{1}{\Phi} = \frac{1}{2\phi_1}\left(1 + \frac{k_3[O_2][M]}{k_2[O_3]}\right) \tag{1.27}$$

Figure 1.2 shows $1/\Phi$ plotted against $[O_2][M]/[O_3]$ for some real experimental results on ozone photolysis by red light. The graph is sensibly linear over the entire range, which provides strong, but circumstantial, evidence that the hypothetical mechanism is correct. An intercept $(=1/2\phi_1)$ of very nearly 0.5 is obtained, and the primary quantum yield is therefore close to unity. Primary processes other than decomposition in reaction (1.17) need not be sought. The slope of the graph suggests that, at $18\,^\circ\text{C}$, $k_3/k_2 \simeq 50 \text{ dm}^3 \text{ mol}^{-1}$, a result in excellent agreement with the values of k_2 and k_3 quoted earlier and obtained by independent absolute techniques.

One feature of steady-state experiments, which is highlighted by the example, is that only *ratios* of rate constants can be obtained from the kinetic analysis. This result is a consequence of the steady state being established because of competition between production and loss processes. Although much valuable information can be extracted, especially if one of the rate constants involved can be determined absolutely in some other way, the limitation does seriously restrict the applicability of steady-state experiments to kinetic investigations. The alternative approach is to study the reaction system in a *time-resolved* manner, using *non-stationary* conditions. Given that, for highly reactive intermediates, the steady state is reached in very short time

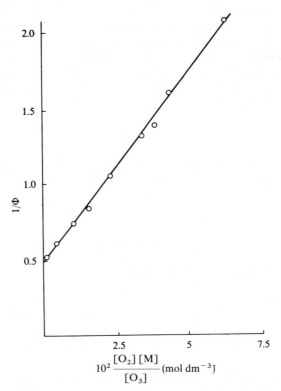

Fig. 1.2 Plot of $1/\Phi$ against $[O_2][M]/[O_3]$ for ozone photolysis by red light. (From Castellano, E. and Schumacher, H. J. (1962). *Z. Phys. Chem.* (*Frankfurt am Main*) **34**, 198.)

periods, the non-stationary experiments have required the development of suitable techniques with the appropriate time resolution. In photochemistry, one of the most powerful and important of these techniques has been *flash photolysis* (cf. Section 7.5). A short-duration flash of light from a discharge tube or laser is used to initiate the photochemical reaction, and the transient changes in concentration of reactants, products, or most frequently, the intermediates themselves are monitored as a function of time. Early experiments were performed on the millisecond time-scale—a major advance at the time—and further developments have made possible the study of processes occurring on the microsecond, nanosecond, picosecond, and even femtosecond (10^{-15} s) time-scales. Discussion of the method itself belongs in Chapter 7, but the kinetic aspects of non-stationary experiments are well illustrated by the results of flash experiments. Let us revert to the example of ozone photolysis, and consider what would happen if the ozone were exposed to a flash of light rather than to steady illumination. Immediately after the flash, some of the ozone would have been dissociated to form atomic oxygen. Suppose the concentration of O to be $[O]_0$ just after the flash has finished. No further O will be formed, but it will be lost in reactions (1.18) and (1.19), so that a simplified form of eqn (1.22) describes the loss of O:

$$\frac{d[O]}{dt} = -k't \tag{1.28}$$

So long as the quantity of ozone that is lost in the secondary reactions is small, k' can be assumed to be roughly constant, and the differential equation can be integrated to yield

$$[O] = [O]_0 \exp(-k't)$$
$$= [O]_0 \exp\{-(k_2[O_3] + k_3[O_2][M])t\} \tag{1.29}$$

A plot of $\ln[O]$ against time will therefore have a slope of $-(k_2[O_3] + k_3[O_2][M])$. Because the plot is logarithmic, relative atomic concentrations will suffice. But the most significant feature of eqn (1.29) is that the rate constants k_2 and k_3 are separated out as a sum, rather than combined in a ratio as in the steady-state expression (1.27). It is therefore easy to design an experiment in which values of the variable k' are obtained as a function of $[O_3]$, say, to extract an absolute value for k_2.

It will be seen that the preceding discussion was based on measurements of (relative) concentrations of the reaction intermediate, atomic oxygen. In fact, measurements of $[O_3]$ decay could also yield absolute rate constants, although the mathematical expressions are more complex. However, in reality, much modern work on photochemical kinetics does rely on direct investigation of the reaction intermediates. Flash and related methods lend themselves particularly well to such an approach, because the peak concentrations of intermediates is much higher than the steady-state concentrations obtained with continuous illumination. A wide range of techniques is available for the study of the intermediates, including several variants of optical spectroscopy in emission and absorption, electron paramagnetic resonance spectroscopy, and mass spectrometry. The subject is a fascinating and important one in its own right, and we shall refer to some of the techniques as the occasion may demand, and in greater detail in Section 7.4.

In this section, we have seen how an analysis of reaction rate data can lead to a better understanding of primary and secondary photochemical processes. Rate data afford valuable insights into the mechanisms of reactions and even into the detailed nature of individual elementary reaction steps. The kinetic approach is a valuable adjunct to studies of absorption spectra, fluorescence, and many other optical and photochemical phenomena, and its use will be implicit in many of the discussions of the following chapters.

Bibliography

Some general textbooks on photochemistry that contain good introductory material:

Calvert, J. G. and Pitts, J. N., Jr (1966). *Photochemistry*. John Wiley, Chichester and New York.

Okabe, H. (1978). *Photochemistry of small molecules*. John Wiley, Chichester and New York.

Barltrop, J. A. and Coyle, J. D. (1975). *Excited states in organic chemistry*. John Wiley, Chichester and New York.

Barltrop, J. A. and Coyle, J. D. (1978). *Principles of photochemistry*. John Wiley, Chichester and New York.

Turro, N. J. (1978). *Modern molecular photochemistry*. Benjamin/Cummings, Menlo Park, CA.

Horspool, W. M. (1976). *Aspects of organic photochemistry*. Academic Press, New York.

Cowan, D. O. and Drisko, R. L. (1976). *Elements of organic photochemistry*. Plenum Press, New York.

Cox, A. and Kemp, T. J. (1971). *Introductory photochemistry*. McGraw-Hill, New York.

Coyle, J. D. (1986). *Introduction to organic photochemistry*. John Wiley, Chichester and New York.

Relevant articles:

Oster, G. (1968). The chemical effects of light. *Sci. Am.* **219**(3), 158.

Dauben, W. G., Salem, L., and Turro, N. J. (1975). A classification of photochemical reactions. *Acc. Chem. Res.* **8**, 41.

Older general accounts of historical importance:

Bowen, E. J. (1946). *Chemical aspects of light*. Oxford University Press, Oxford.

Noyes, W. A. and Leighton, P. A. (1941). *The photochemistry of gases*. Prentice-Hall, Englewood Cliffs, NJ.

2

Absorption and emission of radiation

2.1 Introduction

Photochemistry is intimately dependent on processes involving the absorption or emission of radiation, and it seems desirable to provide at least a brief summary of these processes. Detailed description of spectroscopic phenomena would, however, be out of place, and for amplification of the remarks offered here the reader should refer to one of the texts listed in the Bibliography at the end of the chapter. The survey of this chapter is intended only to serve as a reminder of those parts of spectroscopy needed later in the book.

Spectroscopic nomenclature is often used by photochemists to denote a specific electronic state of some species. Such nomenclature may be the most convenient way of defining a state and distinguishing it from others, although it may sometimes be unnecessarily precise. Those not familiar with the terminology may stop to unravel the meaning of a term symbol when all that had been intended by the writer was to denote that some chemical species was (or was not) excited. A section (2.5) of this chapter briefly sets out the terminology adopted in this book.

2.2 Electromagnetic radiation

An appreciation of absorption or emission processes requires some understanding of the nature of light. The particular question that we have to ask ourselves as photochemists is how light can alter the electronic configuration in an absorbing species, or how a change in the configuration can lead to emission of light.

From the time of Newton until the advent of quantum theory, the corpuscular (or 'particle') theory of light lost ground to the wave theory. Phenomena such as diffraction, or more especially interference, were only explicable in terms of a wave theory. However, the actual nature of the wave, and the mechanism of its propagation, was not established until the latter part of the nineteenth century. In the 1860s, James Clerk Maxwell made one of the major contributions to physics: possibly the only earlier work of such stature was that of Newton. Maxwell was attempting to reconcile the laws of electricity with those of magnetism. By powerful mathematical reasoning, Maxwell demonstrated that such reconciliation would be possible if, associated with an oscillating magnetic field, there were a similar electric field, and vice versa, and *if a wave were propagated in a direction perpendicular to a plane containing the electric and magnetic fields.* The derivation of Maxwell's equations need not concern us here, but one feature of the equations is of the greatest importance. The velocity of propagation of Maxwell's 'electromagnetic' waves *in vacuo* was shown to be numerically identical to the velocity of light *in vacuo*, as determined by Römer (1675), Fizeau (1849) or Foucault (1862). This striking result (1865) obviously suggests that light is an electromagnetic wave, but it did not draw much attention until after Hertz had confirmed (1887–8) Maxwell's prediction of propagated waves from systems involving oscillating electrical and magnetic fields.

The events leading to the awareness that light is a form of electromagnetic radiation have been emphasized here, since a scientist of the second half of the twentieth century has, as part of his 'culture', the belief that light is a form of electromagnetic radiation. We also understand that radio waves, infrared radiation, X-rays and cosmic rays, as well as light and ultraviolet radiation, are electromagnetic radiation, and that they differ from one another only in terms of their frequencies. The most significant modification of Maxwell's nineteenth-century picture of electromagnetic radiation is our awareness that wave motion may have particulate properties associated with it, and that the energy of the particle, or photon, ε, and frequency, v, of the wave are related by $\varepsilon = hv$ (see Section 1.2).

Maxwell's electromagnetic field theory describes radiation in terms of oscillating electric and magnetic fields. It is one or other of these fields (usually the electric one) that interacts with the electrons of the chemical species absorbing the radiation.

2.3 Absorption and emission processes

In this section an attempt is made to describe absorption and emission processes in terms of mechanical models. A more complete and satisfactory representation of the processes is given mathematically by solutions of the *time-dependent Schrödinger equation*. Presentation of the mathematical equipment needed to deal with this approach would, however, obscure the basic principles that we wish to develop; where necessary, the results of the wave-mechanical reasoning are given without proof.

There are three processes that we must distinguish: *absorption, stimulated emission,* and *spontaneous emission.* Suppose a chemical species possesses two quantized states l and m, of energies ε_l and ε_m. If the species is in state l initially, it might be able to interact in some way with electromagnetic radiation and absorb energy in order to reach state m. In a normal process this absorption of energy occurs in a single step so that the energy difference between final and initial levels must be equivalent to the energy of a single photon of radiation. Hence, absorption of radiation can only occur if $\varepsilon_m - \varepsilon_l = h\nu$ ('*Bohr condition*'). The process of *absorption* has involved the loss of intensity from the electromagnetic radiation and the gain of energy by the absorbing species. The converse process, in which a species in an upper state gives up energy to electromagnetic radiation and increases the intensity, is known as *stimulated emission*: the word 'stimulated' indicates that it is the interaction between the radiation already present and the energy-rich species that encourages the latter to give up its energy. Although we have not mentioned the nature or magnitude of the interaction between the species and radiation, it is apparent that the rate (intensity) of absorption or stimulated emission is proportional to the rate of 'collision' between photons and the absorber or emitter: that is to say, the intensity change is proportional to the radiation density, ρ, and to the concentration of chemical species. The constant of proportionality defines the so-called Einstein '*B*' coefficients. B_{lm} is the coefficient for absorption, while B_{ml} is that for stimulated emission: the principle of microscopic reversibility suggests that $B_{lm} = B_{ml}$, and this result can also be derived from the complete treatment of radiation theory. The rates of absorption and stimulated emission are $B_{lm}n_l\rho$ and $B_{ml}n_m\rho (= B_{lm}n_m\rho)$, respectively, where n_l and n_m are the concentrations of species in lower and upper states. For a system in thermal equilibrium, n_m is always less than n_l (see the Boltzmann equation 1.4) and absorption is always a more important process than stimulated emission. How much more important depends, of course, on the relation between $(\varepsilon_m - \varepsilon_l)$ and the temperature, T. It has been pointed out that the energy levels of significance in photochemistry are such that $(\varepsilon_m - \varepsilon_l) \gg kT$, and $n_m \ll n_l$, so that stimulated emission is rarely important in photochemical processes in which thermal equilibrium is established. However, in non-equilibrium situations, stimulated emission may not be negligible, and if a *population inversion* $(n_m > n_l)$ arises, then the emission process will predominate over absorption, and net emission will result. The LASER (Light Amplification by Stimulated Emission of Radiation) depends on the achievement of such population inversions, generally by photochemical techniques (see Section 5.7).

In addition to absorption and stimulated emission, a third process, *spontaneous emission*, is required in the theory of radiation. In this process, an excited species may lose energy in the absence of a radiation field to reach a lower energy state. Spontaneous emission is a random process, and the rate of loss of excited species by spontaneous emission (from a statistically large number of excited species) is kinetically first-order. A first-order rate constant may therefore be used to describe the intensity of spontaneous emission: this constant is the Einstein '*A*' factor, A_{ml}, which corresponds for the spontaneous process to the second-order B constant of the induced processes. The rate of spontaneous emission is equal to $A_{ml}n_m$, and intensities of spontaneous emission can be used to calculate n_m if A_{ml} is known. Most of the emission phenomena with which we are concerned in photochemistry — fluorescence, phosphorescence, and chemiluminescence — are normally spontaneous, and the descriptive adjective will be dropped henceforth. Where emission is stimulated, the fact will be stated.

We referred in the last paragraph to the calculation of concentrations of excited species from emission intensity measurements. It may, however, not always be possible to determine A_{ml} directly by the techniques to be discussed in Chapter 7, and some other method of evaluating A factors may be needed. The A coefficient may be calculated from the B coefficient for the same transition by using the relation

$$A_{ml} = \frac{8\pi h v^3}{c^3} B_{lm} \tag{2.1}$$

B itself may be determined from experimental absorption measurements, as will be described in Section 2.4. The equation is the 'v^3 law' to which reference is frequently made by spectroscopists and photochemists.

The nature of the interaction between electromagnetic radiation and matter must now be considered. The processes may become clearer if we consider a simple example: the absorption of infrared radiation by a molecule of HCl. The molecule has a permanent dipole moment, so that the energy of the molecule will be affected by the presence of an electric field, and the bond will tend to be distorted according to the direction of the field. Now consider an oscillating electric field, such as that present in electromagnetic radiation. If the frequency of oscillation is equal to the vibration frequency of the H–Cl bond, then the induced motion of the electrons may lead to an increased energy of nuclear motion. The vibrational energy in the molecule will then increase by one quantum, and the intensity of electromagnetic radiation will be depleted by an equivalent amount. (The basic tenet of the quantum theory is that the distortions occur *only* in the quantized units, which will lead to absorption of a whole photon.) This description of the absorption process is obviously just a pictorial representation, but it indicates that the interaction derives from the influence, via the molecular dipole, of the electric vector of the radiation on the energy of the molecule. A transition occurring through such an interaction is called an *electric dipole transition*. Interactions with the magnetic vector of the radiation, or those that involve quadrupoles in the chemical species, give rise to *magnetic dipole*, *electric quadrupole*, and *magnetic quadrupole* transitions, etc. All these interactions are, however, much weaker than the electric dipole interaction and may frequently, but by no means always, be ignored.

The absorption or emission of infrared radiation by an oscillating molecule possessing a dipole is readily understood in the pictorial terms of the last paragraph. It is less easy to describe electronic transitions in the same manner. In the classical sense, electronic excitation does not correspond to increasing the energy in an oscillating system, and, in any case, neither upper nor lower electronic state may possess a steady dipole (e.g. the electron cloud in an atom is symmetrically disposed about the nucleus in all states, so there is no effective charge separation). However, the general principles of interaction with radiation still apply, and what we need to know is whether an (electric) dipole interaction can occur during transitions between two states. Wave-mechanical techniques provide the only satisfactory method of dealing with this problem: the time-dependent Schrödinger equation, referred to at the beginning of the section, can be used to derive the rate at which a system can be changed from one stationary state to another under the influence of a perturbing effect. If this rate is non-zero for a perturbation of the system involving electric dipole interaction with the electric vector of radiation, then an electric dipole transition can occur. The rate of change between states multiplied by the number of species present in the lower state is, of course, the overall rate of absorption of photons, so that, in principle, solution of the time-dependent Schrödinger equation should lead to prediction of the intensity of the transition. Explicit solutions are, however, rarely available, and in such cases it may be possible only to say whether or not any interaction occurs, rather than to calculate its magnitude. The conditions under which a specific interaction arises are given as *selection rules* for that type of transition: for electric dipole transitions in electronic spectroscopy, they are given in Section 6 of this chapter.

In this section we have distinguished between spontaneous and induced transitions, and we have shown how the probabilities for these processes, the Einstein A and B coefficients, are related to each other. The following section deals with experimental measurements of absorption, and the relation between these measurements and the theoretical quantities is explored.

2.4 Absorption of radiation: the Beer–Lambert law

The fraction of light transmitted through an absorbing system is very frequently found to be represented by the relation

$$\frac{I_t}{I_0} = 10^{-\varepsilon C d} \tag{2.2}$$

I_t, I_0 are transmitted and incident light intensities, C is the concentration of absorber, d is the depth of absorber through which the light beam has passed, and ε is a constant of proportionality known as the *decadic extinction coefficient*[†], which is dependent on the wavelength of radiation (and may occasionally vary with C — this question is mentioned later). The law embodied in eqn (2.2) was originally known as Lambert's law; a second law, Beer's law, stated that if C and d were altered but the product Cd was constant, then the fraction of light transmitted remained the same. Since this latter law follows in any case from Lambert's law, the relation (2.2) is now known as the *Beer–Lambert law*. The logarithmic form of the equation

$$\log_{10} I_0/I_t = \varepsilon Cd \tag{2.3}$$

is often employed, and the product εCd is called the *Optical Density* (O.D.) of the system.

A 'proof' of the Beer–Lambert law may be derived if it is assumed that the rate of loss of photons is proportional to the rate of bimolecular collisions between photons and the absorbing species. The decrease, $-\mathrm{d}I$, in intensity I at any point x in the system (see Fig. 2.1) for a small increase in x, $\mathrm{d}x$, is given by

$$-\mathrm{d}I = \alpha I C\,\mathrm{d}x \tag{2.4}$$

where α is a constant of proportionality. Integration, with the boundary conditions $I=I_0$ at $x=0$, $I=I_t$ at $x=d$, yields the result

$$\frac{I_t}{I_0} = \mathrm{e}^{-\alpha Cd} \tag{2.5}$$

which is the same as eqn (2.2) with $\alpha = 2.303\varepsilon$.

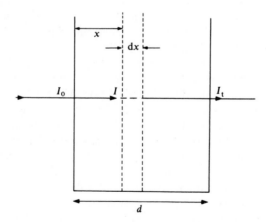

Fig. 2.1   The change in intensity, I, with optical path, x (see eqns 2.4 and 2.5).

[†] The term *extinction coefficient* is normally employed, although IUPAC recommends that *absorption coefficient* be adopted. Both terms are in common use.

Both natural and decadic extinction coefficients are used in practice, and it is essential to state the base as well as the units of C in referring to extinction coefficients (the units of d are almost always cm). Thus a 'decadic molar extinction coefficient' is ε of eqn (2.2) with C in $mol\,dm^{-3}$ and d in cm.

The intensity of radiation absorbed, I_{abs}, is, of course, $I_0 - I_t$, so that the fraction absorbed is given by

$$\frac{I_{abs}}{I_0} = 1 - 10^{-\varepsilon Cd} = 1 - e^{-\alpha Cd} \tag{2.6}$$

An important approximate expression results when αCd is small: expansion of the exponential $e^{-\alpha Cd}$ and rejection of second- and higher-order terms in αCd leads to the conclusion that

$$\frac{I_{abs}}{I_0} \sim \alpha Cd \tag{2.7}$$

For $\alpha Cd = 0.01$, the approximate value of I_{abs}/I_0 differs from the accurate one by $< 1\%$, while even for $\alpha Cd = 0.1$, the difference is only 5%. Thus, eqn (2.7) often gives a sufficiently accurate estimate of fractional absorptions less than about 10%.

Although the Beer–Lambert law usually offers an adequate description of experimental data, there are some circumstances in which it does not. For example, the width of an absorption band or line depends, in part, on factors such as molecular collision, so that changes in concentration can alter the ε–λ relationship and hence lead to a breakdown of the law. For oriented systems (e.g. crystals) the value of ε may depend on the plane of polarization of the light. Again, the law holds only if the wavelength range to which the intensity measurements refer is small compared with the width of the absorption band (i.e. if ε is constant over the wavelength range). Thus, experiments in which there is absorption from wide-band incident radiation by a narrow-band absorber will not obey the Beer–Lambert law. More trivially, the concentrations of species that associate or dissociate will not be equal to the concentrations expected on the basis of amounts of material added.

As we shall see later in the book, the maximum value achieved in an absorption band by the extinction coefficient may be an indicator of the nature of the spectroscopic transition, and especially of whether the transition is 'allowed' or 'forbidden' for electric dipole interactions. Although the experimental extinction coefficients α or ε are measured, in principle, at a single wavelength or frequency, a real absorption band spans a range of frequencies. The experimental measure of the probability of the interaction is the absorption coefficient integrated over the band, $\int \alpha d\nu$, rather than the absorption coefficient at a single frequency, the limits of integration being set by the frequency limits of the transition under consideration. The Einstein B

coefficient (see Section 2.3) is one measure of the probability that a transition will occur somewhere in the band, and is connected to the integrated absorption coefficient by the relation

$$B_{lm} = \frac{c}{hv} \int_{\text{band}} \alpha \, \mathrm{d}v \qquad (2.8)$$

This equation provides the means of estimating B (and hence A: cf. eqn (2.1)) from absorption measurements. For a given overall probability of transition, there is some kind of inverse relationship between the band width and the extinction coefficient. However, for a typical band width, the maximum value of the decadic molar extinction coefficient rarely exceeds 10^5 cm^{-1} dm^3 mol^{-1}: a value of 5×10^4 cm^{-1} dm^3 mol^{-1} is considered usual.

2.5 Spectroscopic nomenclature

The spectroscopic nomenclature used in later parts of the book needs to be explained, although it is out of place to show how, for example, term symbols for atoms are derived. It seems more pertinent to the present purpose to explain the function of the nomenclature, and then to cite briefly the actual forms used in the book.

For the spectroscopist, the object of the nomenclature is to define the electronic arrangement—or electronic 'state'—of the two levels involved in a spectroscopic transition; for the photochemist, the object may be to specify the excited state implicated in a photochemical process. An extremely important quantity in defining the electronic state is the total spin of all the electrons, S. This spin is obtained by appropriate vector summation of the electron spin moments of each individual electron (equal to a half-unit). It is common practice to specify the *spin multiplicity* of a species rather than the spin itself. The multiplicity is the degeneracy, $2S+1$, that might be lifted, for example, by application of a magnetic field. For species in which all the electrons are paired, $S=0$, $2S+1=1$ and the species is a *singlet*. Most chemical species that can be stored for long periods are of this kind, although there are exceptions (NO has $S=\frac{1}{2}$, O_2 has $S=1$). If two electrons are unpaired with parallel spins, $S=1$, $2S+1=3$, and the species is in a *triplet* state. Triplet excited states are obviously possible for species with singlet ground states. Free radicals such as OH or ClO have one unpaired spin, so that $S=\frac{1}{2}$, $2S+1=2$, and they are *doublets*. The special significance of spin is that it cannot be altered in interactions with radiation (or, for that matter, in a radiationless transition—see Fig. 1.1, pathway v). States of the same chemical species, but possessing different multiplicities, tend to behave as distinct groups with little propensity to convert from one to the other. Certainly, the groups of states *do* interconvert, as we shall see later, because spin is not as 'pure' a quantity as these introductory sentences suggest. Nevertheless, it remains true that both radiative and radiationless transitions are much less efficient between states of different multiplicity than they are between states of the same multiplicity.

The observations concerning spin multiplicity lead to one of the simplest methods of describing states. If all that matters is the energy order of the states and their multiplicities, a numbered list based on the multiplicity may be drawn up. In the common case where the ground state is a singlet, this state is called S_0, and the excited singlets written as S_1, S_2, etc., in order of their energies. The triplet states are T_1, T_2, etc., T_0 being excluded because it is not the ground state.

A more detailed description of electronic states is sometimes possible, and it may be necessary to invoke it to specify a state unambiguously. For example, the ground state of the Na atom may be represented as $1s^2 2s^2 2p^6 3s^1$ and the first electronically excited state as $1s^2 2s^2 2p^6 3p^1$, and the photochemist who wishes merely to know the extent and type of excitation might be satisfied by these descriptions. However, transitions between these 'two' levels give rise to the well-known *doublet* yellow D lines; as a result of differing interactions between electron spin and orbital momentum, there are two different states, of different energies, although both are described by $1s^2 2s^2 2p^6 3p^1$. If, then, a closer definition of the state is required, some further information must be provided. This information is contained within the spectroscopic *term symbol*. The two states corresponding to the description $1s^2 2s^2 2p^6 3p^1$ are given the term symbols $^2P_{1/2}$, $^2P_{3/2}$ (or, since both these term symbols can arise from *any* atom having one electron in a p orbital, the principal quantum number may also be stated, and the term symbols written $3(^2P_{1/2})$, $3(^2P_{3/2})$). It is the meaning of the components of the term symbol with which we are primarily concerned. In atomic term symbols such as $^2P_{1/2}$, the superscript number, subscript number and the capital letter each defines a particular angular momentum of the electronic state, as will be described in the next section. It is frequently the *symmetry* properties of two wave functions that determine whether or not an interaction with electromagnetic radiation is possible, so that from the spectroscopist's point of view, symmetry and energy may be the most important features of an electronic state. As it happens, the angular momentum quantum numbers provide some of the symmetry information; and for atoms or for some molecules, a statement of angular momentum quantum numbers forms at least part of the spectroscopic nomenclature.

2.5.1 *Atoms*

Individual electrons possess not only *spin angular momentum*, but they may possess also *orbital angular momentum*. These momenta are vector quantities, and are given the symbols **s** and **l**. Associated with an atom are angular momenta resulting from the vector addition of the **l** and **s** quantities for each electron, a process referred to as *coupling*. The concept of adding spin angular momenta has already been introduced in describing the total spin, S, of an atom or molecule. In one simple scheme of coupling angular momenta, known as *Russell–Saunders coupling*, the individual spin momenta s_1, s_2, etc., couple to give S, while l_1, l_2, etc., couple to give an overall orbital momentum L. Since S and L are themselves momenta, they also have a resultant, the *total angular momentum*, J. Such coupling describes the situation fairly adequately for light atoms, but not for heavy ones; however, Russell–Saunders term symbols are often used for heavy atoms, and the breakdown of selection rules based on S, L, and J (see Section 2.6) reflects the inadequacy of the description.

Each angular momentum, S, L, and, in turn, J, is calculated according to the ordinary laws of vector addition, but with the additional restriction for quantized momenta that addition can take place only in quantized units. For example, J can take the values $L+S$, $L+S-1$, $L+S-2$, ..., $L-S+2$, $L-S+1$, $L-S$ (if $L>S$). Note that there are $2S+1$ values of J, so that the multiplicity introduced earlier is seen to provide an indication of how many J subcomponents might exist for a particular state.

The atomic term symbol is constructed in the form

$$^{2S+1}L_J$$

In order to avoid writing out the number for L, we represent $L=0, 1, 2, 3$, etc., by the capital letters S, P, D, F, etc., following directly the convention for orbital notation that electrons with $l=0, 1, 2, 3$ should be called s, p, d, f electrons. For example, the sodium atom has only one electron of interest, and $L=l$, $S=s$. In the first excited state ($\ldots 3p^1$), $L=1$, $S=\frac{1}{2}$, and, therefore, $J=\frac{1}{2}$ or $\frac{3}{2}$. Hence, the states are $^2P_{1/2}$, $^2P_{3/2}$, as given earlier. In speech, the states are called 'doublet-P-one-half' and 'doublet-P-three-halves', the states with multiplicities 1, 2, 3, etc. originally being known as singlets, doublets, triplets, etc. after the appearance of spectra involving them.

2.5.2 *Diatomic and linear polyatomic molecules*

In diatomic and linear polyatomic molecules, the electronic state may still be defined in part by the magnitude of the orbital electronic angular momentum resolved along the internuclear axis. The nomenclature follows the general pattern, with Roman letters replaced by Greek characters. Thus l and L for atoms are replaced by λ and Λ for individual orbitals and the whole molecule, respectively, and orbitals with $\lambda=0, 1, 2, 3$ are the familiar σ, π, δ, φ orbitals, while states with $\Lambda=0, 1, 2, 3$ are Σ, Π, Δ, Φ states. The term symbol consists basically of

$$^{2S+1}\Lambda$$

although a number of other pieces of information may be added on. One of these is, of course, the total angular momentum, and one of several possible coupling schemes must be used to derive it. Secondly, some further description of the symmetry properties of the wave function may be possible. A wave function may possess some, or all, of the molecular symmetry. In particular, for a centrosymmetric molecule, the wave function may either remain unchanged or change sign (but not magnitude) on inversion through the centre of symmetry. Such wave functions are called 'even' or 'odd', respectively: in German the words are 'gerade' and 'ungerade', and the symbols g or u, given as subscripts after Λ, derive from the initial letters. The wave function for a Σ state ($\Lambda = 0$) may remain the same or change sign on reflection by a plane of symmetry passing through the line of atomic centres: these two possibilities are represented by the symbols $+$ or $-$ appearing as superscripts after Λ. A few examples will illustrate the forms used.

N_2—the ground state is $^1\Sigma_g^+$: there is neither spin nor orbital angular momentum since all orbitals are closed (fully occupied), and the wave function does not change sign either on inversion through a centre of symmetry or on reflection.

CO_2—the ground state is $^1\Sigma_g^+$: the same remarks apply to this linear triatomic molecule as to N_2.

N_2O—the ground state is $^1\Sigma^+$: since the molecule has no centre of symmetry, the wave function cannot be either g or u.

NO—the ground state is $^2\Pi$: there is one electron in an unfilled orbital, which is an antibonding π orbital. S is therefore $\frac{1}{2}$, $(2S+1)=2$; Λ is 1, and the molecule is in a Π state. NO does not possess a centre of symmetry.

O_2—the ground state is $^3\Sigma_g^-$: there are two electrons in the antibonding π orbitals, which can give rise to $S=0$ or 1 and $\Lambda=0$, 1, or 2. Not all the apparent combinations are allowed because of the operation of the exclusion principle. $^3\Sigma_g^-$ is the lowest energy arrangement, while the states $^1\Delta_g$ and $^3\Sigma_g^+$ (among others) are of rather higher energy. Note that several energy levels can derive from different coupling of electrons, without moving an electron to a different orbital.

2.5.3 *Small, non-linear molecules*

It is not practicable to specify the electronic state of a non-linear molecule in terms of orbital angular momentum, although spin multiplicity is still meaningful. If the molecule possesses symmetry elements, and the electrons are still sufficiently delocalized for the electronic cloud to possess effectively the same symmetry, then it may be possible to describe the electronic state in terms of the effect that the symmetry operations have on the sign of the wave function. Symmetry symbols such as A, B, E, and T, with various subscripts and superscripts, are used to classify the behaviour of the wavefunction under the various symmetry operations. As usual, the symbol is preceded by a superscript number giving the spin multiplicity. The numerical subscript may be followed by a g or u if the molecule has a centre of symmetry. The symbols are, in fact, the labels of the rows of the 'character tables' of group theory, but that does not directly concern us here. It is, however, worth pointing out that a wavefunction of the type A, A_1, or A' is symmetric with respect to all the molecular symmetry operations, while those of the type A_2 or A" are antisymmetric with respect to planes of symmetry present in the molecule. The various B types indicate antisymmetry with respect to rotational operations as well. Typical examples of term symbols for the ground states of simple molecules would be $H_2CO(^1A_1)$, $NO_2(^2A_1)$, and $C_6H_6(^1A_{1g})$; one excited state in each case is $H_2CO(^1A")$, $NO_2(^2B_1)$, and $C_6H_6(^1B_{2u})$. Note that the change from subscript number to superscript primes in the case of excited H_2CO indicates that the symmetry of the molecule has changed slightly on excitation (the molecule becomes slightly non-planar).

2.5.4 *Complex molecules*

Complex molecules may not possess any symmetry elements, or if they do, the localizations of the electrons can so distort the electron cloud that its symmetry bears little relation to the molecular symmetry. In such cases it may be best to revert to a description of states in terms of the individual orbitals. As an example, we will consider formaldehyde, although a molecule as simple as this is probably best described by the group-theoretical term symbol of the last paragraph. The last filled orbitals in H_2CO can easily be shown to be . . . $(\pi_{CO})^2 (n_O)^2$, where n_O represents the non-bonding orbital on the O atom and the two electrons in it are the 'lone pair'. The first unfilled orbitals in formaldehyde are the π_{CO}^* and σ_{CO}^* antibonding orbitals. Promotion of one electron from, say, the n_O orbital to π_{CO}^* leads to excitation, and there is no restriction now on unpairing of spins so both singlet and triplet states are possible. The states are then designated as $^3(n, \pi^*)$ or $T_1(n, \pi^*)$, $^1(n, \pi^*)$ or $S_1(n, \pi^*)$, and similarly for (π, π^*), (n, σ^*), and (π, σ^*), states. Spectroscopic transitions are then referred to as $n \rightarrow \pi^*$, and so on.

2.6 Selection rules for optical transitions

Formal rules, known as *selection rules*, may be used to decide whether or not an electric dipole transition between two states may take place. Since these rules will be alluded to at several points later in the book, it seems desirable to list the most frequently met cases at this stage.

Perhaps the most important is the rule governing spin multiplicity: spin must not change during an electronic transition. The usual way to write rules of this kind is

$$\Delta S = 0 \tag{2.9}$$

In atoms, for one-electron transitions, we have the selection rules

$$\Delta \mathbf{L} = \pm 1 \tag{2.10}$$

$$\Delta \mathbf{J} = 0, \pm 1, \text{ but } \mathbf{J} = 0 \nleftrightarrow \mathbf{J} = 0 \tag{2.11}$$

For diatomic and linear polyatomic molecules the orbital momentum rule is

$$\Delta \Lambda = 0, \pm 1 \tag{2.12}$$

and, where applicable, the rules governing symmetry

$$u \leftrightarrow g, \; + \leftrightarrow +, \; - \leftrightarrow - \tag{2.13}$$

The rules are derived, as indicated in Section 2.3, by considering whether or not an interaction with the electric vector of electromagnetic radiation is possible in going from the initial to the proposed final states of the chemical species. Frequently, the symmetry properties of the wavefunctions involved suffice to exclude certain transitions, so that the selection rules are a way of representing the possibilities that remain, although they say nothing about the absolute intensities of the interactions.

For non-linear molecules that still possess some symmetry, selection rules can again be derived on the basis of the symmetry properties. If the molecule possesses a centre of symmetry, then g and u states exist, and electric dipole transitions require that $g \leftrightarrow u$. Molecules with symmetry of ground state H_2CO or NO_2 have symmetry-allowed transitions $A_1 \leftrightarrow B_1, B_2$, but not from A_1 to A_2, or B_1 to B_2. Similarly, molecules with the symmetry of C_6H_6 cannot undergo symmetry-allowed electric dipole transitions from the A_1 ground state to a B_1, B_2, or any other A_1 state, and since the molecule is centrosymmetric, transitions from the g ground state to any other g state are also excluded.

Although the selection rules provide a useful guide to the types of transition that might be expected, they are by no means followed rigidly. Absorption or emission spectra due to 'forbidden' transitions are quite often observed, although they are usually considerably weaker than the 'allowed' transitions. An example that follows directly from the preceding paragraph concerns the $^1B_{2u} \leftarrow {}^1A_{1g}$ absorption in benzene. According to the rules propounded, the transition is forbidden. However, the symmetry rule is based on the shape of benzene being a perfect hexagon. In reality, molecular vibrations distort the molecule so that the symmetry is reduced and the transition becomes weakly allowed. Such coupling of vibrational and electronic motions (*vibronic coupling*) is a consequence of a breakdown in the Born–Oppenheimer approximation, which states that all energy modes (e.g. electronic, vibrational, and rotational) can be considered independently, so that the individual wavefunctions can be factorized. Breakdown of the separability of *rotational* and electronic modes is much less important, although it is still recognized as a source of very weak forbidden transitions.

The spin selection rule, $\Delta S = 0$, might be expected to be of universal applicability, since it does not require the molecule under consideration to have any geometrical symmetry. However, 'spin-forbidden' transitions are also frequently observed. The spin rule is based again on the idea of separability of wavefunctions, this time of the spin and spatial components of the electronic wavefunction. However, the electron experiences a magnetic field as a result of the relative motion of the positive nucleus with respect to it, and this field causes some mixing of spatial and spin components, giving rise to *spin–orbit coupling*. As a result, the idea of 'pure' spin states must be modified to allow for exchange of angular momentum with the orbital mode. For example, a state formally described as a singlet may, in reality, have some triplet character, while a formal triplet may have some singlet character. Transitions between singlets and triplets may then be looked on as transitions between the 'pure' singlet and triplet components of the mixed states. Since spin–orbit coupling depends on interaction with the nucleus, its magnitude increases rapidly with increasing nuclear charge (in fact, as the fourth power). Spin-forbidden transitions are thus stronger when 'heavy' nuclei are involved. One important example concerns mercury 'resonance' radiation. (Resonance radiation refers to emission from the first excited state to the ground state: *resonant* absorption and re-emission may occur). The ground state of mercury is $^1S_0$, and the first excited singlet is $^1P_1$. Transitions of the $^1P_1 \rightarrow {}^1S_0$ line yield the true resonance line at 184.9 nm, and it is so readily reabsorbed by mercury vapour that it is necessary to cool mercury lamps used for its generation. The first excited *triplet* states are $^3P_{0,1,2}$ and transitions of the $^3P_1 \rightarrow {}^1S_0$ line at $\lambda = 253.7$ nm are only weaker by a factor of about a hundred than those of the $\lambda = 184.9$ nm line, even though they are nominally forbidden by the $\Delta S = 0$ selection rule. The intensity of the forbidden line is, in fact, so great that the line at $\lambda = 253.7$ nm is commonly called the resonance line. The explanation for the breakdown of the rule in this instance is that, since mercury is a heavy atom, Russell–Saunders coupling does not really hold. Thus S is not a good quantum number for mercury, and selection rules based on it need not be expected to apply rigorously. For the light element helium, however, which also has a $^1S_0$ ground state and a $^3P_1$ first excited state, the $^3P_1 \rightarrow {}^1S_0$ transition is very many orders of magnitude weaker than the singlet → singlet transition.

There are two other fairly common causes of apparent breakdown of the electronic selection rules. First, collisions with other atoms or molecules, or the presence of electric or magnetic fields, may invalidate selection rules based on state descriptions of the unperturbed species. Secondly, although the transition may be forbidden for an electric dipole interaction, it may be permitted for the (much weaker) magnetic dipole or electric quadrupole transitions.

Not only do the selection rules fail to predict the occurrence of forbidden transitions, but they may also fail to predict that a transition will be weak even though it is symmetry-allowed. One well-known example is that of the $n \rightarrow \pi^*$ transition in pyridine. The transition is of the type $A_1 \rightarrow B_2$, and the transition is symmetry-allowed. However, the non-bonding electron is on the nitrogen atom in C_5H_5N, while the π-orbital involved in the transition is delocalized over the ring. As a result, there is little overlap between the initial and final wavefunctions, and electromagnetic radiation cannot excite the electron from the lower to the upper orbital. The transition is therefore weak (about 100 times less intense than would be expected ordinarily) because of the lack of *orbital overlap*, even though it is allowed on symmetry grounds.

So far, this discussion of selection rules has considered only the electronic component of the transition. For molecular species, vibrational and rotational structure is possible in the spectrum, although for complex molecules, especially in condensed phases where collisional line broadening is important, the rotational lines, and sometimes the vibrational bands, may be too close to be resolved. Where the structure exists, however, certain transitions may be allowed or forbidden by vibrational or rotational selection rules. Such rules once again use the Born–Oppenheimer approximation, and assume that the wavefunctions for the individual modes may be separated. In an unsymmetrical molecule, there are no restrictions on the vibrational transitions that are possible, so that the spectrum is correspondingly complex. In a symmetrical molecule, only vibrational levels of the same vibrational symmetry species in the upper and lower electronic levels can combine with each other. That means that although all symmetric vibrations can combine with each other, for antisymmetric vibrations, only transitions with $\Delta v = 0$, ± 2, ± 4, etc., can occur. Rotational structure in electronic spectroscopy can be particularly complex, because the molecular rotational angular momentum can couple with the electronic angular momentum, several nominal coupling schemes or cases being known. Furthermore, the rotations available to the molecule depend on its shape (linear, symmetric top, etc.), so that it is not profitable to list here the individual rotational selection rules. One familiar example will suffice: for a linear molecule undergoing a $^1\Sigma \leftarrow {}^1\Sigma$ transition, then the selection rules are $\Delta J = 0$, ± 1.

Quite apart from the symmetry-related selection rules, there is one further very important factor that determines the intensity of individual vibrational bands in electronic transitions, and that is the geometries of the two electronic states concerned. The principles involved are illustrated in the next section with reference to the simple example of a diatomic molecule.

2.7 Intensities of vibrational bands in an electronic spectrum

Relative intensities of different vibrational components of an electronic transition are of importance in connection with both absorption and emission processes. The populations of the vibrational levels obviously affect the relative intensities. In addition, electronic transitions between given vibrational levels in upper and lower states have a specific probability, determined in part by the electronic transition probability and in part by the probability of finding a molecule with similar internuclear separations in both states. This last factor is bound up with the *Frank–Condon principle*, which is yet another aspect of the Born–Oppenheimer approximation. The principle states that an electronic transition occurs so rapidly in comparison with vibration frequencies that no change in internuclear separation occurs during the course of a transition. Thus a line depicting a transition on a potential energy diagram must be drawn vertically. The probability of transition occurring in any small range of internuclear distance will then depend on the product of probabilities of a molecule possessing that internuclear distance in each electronic state, and the total transition probability is this probability integrated over all internuclear distances.

Quantum-mechanical treatment of oscillating motion in molecules shows that the internuclear distance–probability function depends on the quantum number v and that the function has $(v + 1)$ maxima and has v nodes between the maxima. Figure 2.2 shows the probability function for a series of levels of a typical anharmonic oscillator: the greater heights of the maxima for large internuclear distance at high values of v correspond classically to the greater time spent by the molecule at the turning point where the restoring force is

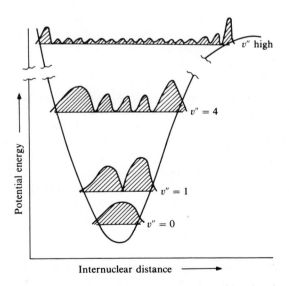

Fig. 2.2 Vibrational probability function for a series of levels of an anharmonic oscillator.

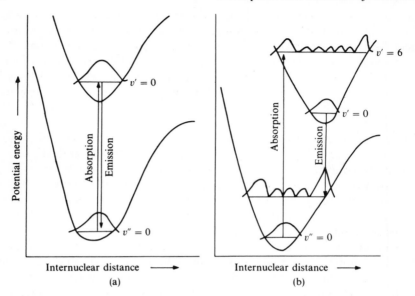

Fig. 2.3   Electronic transitions of greatest probability for absorption and for emission from lowest vibrational levels: (a) where both electronic states are of similar sizes, (b) where the upper state is larger than the lower.

smaller. The relative intensity of transitions can now be predicted by the Franck–Condon principle. In absorption, transitions are most likely to originate from the point of maximum probability in the particular vibrational level of the lower state of the transition, and the relative intensities of transitions to vibrational levels of the upper state will be dependent on the probability of finding the upper state with that internuclear separation. Figure 2.3 illustrates the two different situations that arise (a) when the upper and lower curves are similar in shape and size, and (b) when the upper state is larger than the lower state. In case (a) it can be seen that $(0,0)$ transitions are the strongest, in both absorption and emission, while in case (b) the strongest absorption band is $(6,0)$ and the strongest emission is $(0,4)$. For transitions not involving $v''=0$ (absorption) or $v'=0$ (emission), the $(1,1)$ $(2,2)$, . . . , etc., bands are strong in case (a), and $(5,1)$, $(4,2)$, . . . , etc., are strong in case (b).

Bibliography

Simons, J. P. (1971). *Photochemistry and spectroscopy*. John Wiley, Chichester and New York.

Steinfeld, J. I. (1974). *Molecules and radiation*. Harper and Row, New York.

Hollas, J. M. (1987). *Modern spectroscopy*. John Wiley, Chichester and New York.

Richards, W. G. and Scott, P. R. (1976). *Structure and spectra of atoms.* John Wiley, Chichester and New York.

Richards, W. G. and Scott, P. R. (1985). *Structure and spectra of molecules.* John Wiley, Chichester and New York.

Berkowitz, J. (1979). *Photoabsorption, photoionization and photoelectron spectroscopy.* Academic Press, New York.

Calvert, J. G. and Pitts, J. N., Jr (1966). *Photochemistry*, Chapters 2 and 3. John Wiley, Chichester and New York.

Barltrop, J. A. and Coyle, J. D. (1978). *Principles of photochemistry*, Chapter 2. John Wiley, Chichester and New York.

Herzberg, G. (1950). *Spectra of diatomic molecules.* Van Nostrand, Princeton, NJ.

Herzberg, G. (1966). *Electronic structure and electronic spectra of polyatomic molecules.* Van Nostrand, Princeton, NJ.

3

Photodissociation

3.1 Dissociation as a primary process

An examination of primary photochemical processes may well begin with a discussion of photodissociation, since, of the possible fates of electronically excited species, dissociation most clearly leads to chemical change. The several different paths indicated in Fig. 1.1 are, however, closely interrelated, and in this chapter we shall have to anticipate certain conclusions about fluorescence and energy transfer which are developed more fully in later chapters of the book.

We shall distinguish three major routes to photodissociation—optical dissociation, predissociation, and induced predissociation—and the physical principles involved are described in Sections 3.2–3.4. The illustrative examples chosen in these sections generally involve diatomic inorganic molecules. The potential energy–internuclear distance relationship for a diatomic species is represented in two dimensions by the familiar potential energy curve, and the actual form of the curve is frequently known for specific electronic states of diatomic molecules. Although the same physical principles will apply to the photochemistry of larger molecules, the descriptions are necessarily more complex and less precise. Later sections of the chapter deal with polyatomic molecules of both organic and inorganic species.

3.2 Optical dissociation

An electronically excited species produced by the absorption of light may possess enough, or more than enough, energy to dissociate into fragments. The spectrum of the absorption leading to dissociation is continuous, since the fragments may possess translational energy (which is effectively continuous). At some longer wavelength the spectrum may possibly be banded (although in some cases it is not), in a region where dissociation does not follow absorption. The spectrum of I_2 vapour exhibits typical absorption bands. The bands get progressively closer together until a continuum is reached: the energy corresponding to the onset of the continuum ('convergence limit') is the dissociation energy to the products formed. At room

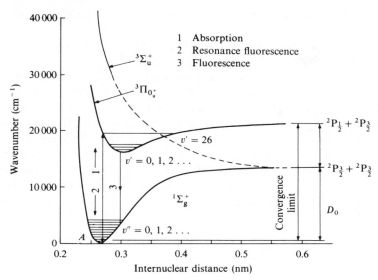

Fig. 3.1 Potential energy curves for the ground state and two excited states of the I_2 molecule. (From Mathieson, L. and Rees, A. L. G. (1956). *J. Chem. Phys.* **25**, 753.)

temperature, almost all iodine molecules are in the ground vibrational level ($v'' = 0$), so that the bands are a progression from $v'' = 0$ and the dissociation energy is the energy from that level. Figure 3.1 gives approximate potential energy curves for the states of I_2 involved (see Section 3.4 for further remarks). The electronic transition $^3\Pi \leftarrow {}^1\Sigma$, although·'forbidden' by the selection rule $\Delta S = 0$, is fairly strong for the heavy molecule I_2 (it becomes progressively weaker for Br_2, Cl_2, and F_2). The potential energy curve indicates that the upper state of I_2 *correlates* (lies on the same curve as) $I(^2P_{1/2}) + I(^2P_{3/2})$: i.e. photodissociation yields one excited ($J = \frac{1}{2}$) as well as one ground-state ($J = \frac{3}{2}$) product atom. One or more excited product fragments are often produced in optical dissociation, and comparison of the spectroscopic convergence limit with the known bond dissociation energy may well reveal the nature of product excitation. For example, in iodine the continuum starts at $\lambda = 498.9$ nm, equivalent to 368 kJ mol^{-1}, while the I–I bond dissociation energy is 151 kJ mol^{-1}. The difference of 217 kJ mol^{-1} corresponds exactly to the excitation energy of the $J = \frac{3}{2}$ to $J = \frac{1}{2}$ state of atomic iodine.

Determination of the precise convergence limit by direct examination of spectra is difficult, and sometimes impossible; a more reliable estimate of the dissociation limit is usually obtained from an extrapolation of measurements taken well into the banded region. Analytical methods may be used occasionally (especially if the potential curve is well represented by the Morse function), but the graphical *Birge–Sponer extrapolation* is of more general application. A comprehensive discussion of the methods is given by Gaydon (Bibliography).

Although the convergence of absorption bands to a continuum is typical of the spectra of small molecules, which are optically dissociated by light absorbed in the continuum, there are some molecules for which the spectrum is continuous over the entire absorption region. A continuous spectrum suggests that either or both of the electronic states involved in the transition are unbound, or so weakly bound that the spacing of vibrational levels is too small to be resolved. It is, of course, improbable that the *lower* state of an ordinary molecule is so weakly bound that it leads to a continuous spectrum, even though some 'quasi' molecules, such as Hg_2 formed in high-pressure mercury vapour, give rise to an absorption continuum. Many molecules do, however, possess repulsive excited states. Figure 3.2 shows some states of

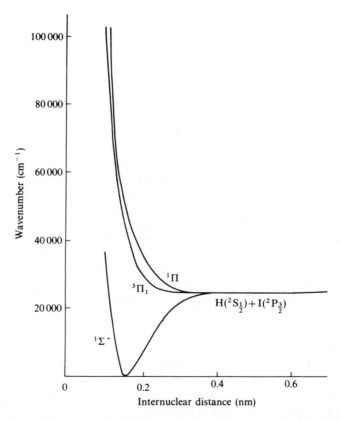

Fig. 3.2   Potential energy curves for some states of HI that correlate with ground-state atoms. (From Mulliken, R. S. (1937). *Phys. Rev.* **51**, 310.)

hydrogen iodide that correlate with ground-state $H(^2S_{1/2})$ and $I(^2P_{3/2})$ atoms; a broad absorption band extends from $\lambda \sim 300$ nm to $\lambda < 180$ nm, and results in part from the $^3\Pi \leftarrow {}^1\Sigma^+$ and $^1\Pi \leftarrow {}^1\Sigma^+$ transitions, and photo-dissociation occurs at all wavelengths in the band. Indeed, since the H–I dissociation energy is about 297 kJ mol^{-1}, there is always considerable excess energy ($\lambda = 300$ nm $\equiv 397$ kJ mol^{-1}) that must be taken up as translation of the product atoms, and there is evidence of increased chemical reactivity of these 'hot' atoms.

Photodetachment of an electron, *photoionization*, can be regarded as a special case of photodissociation, with the ion and electron as dissociation products:

$$AB + h\nu \rightarrow AB^* \rightarrow AB^+ + e \qquad (3.1)$$

Rydberg series (in which the principal quantum number increases) are known for both atoms and molecules, and the lines or bands converge as the electron is moved into orbitals further away from the nucleus. The convergence limit corresponds to complete removal of the electron, and thus to ionization. Experiments using reaction vessels with windows are limited to wavelengths longer than about 105 nm (the cut-off of lithium fluoride: cf. footnote on p. 155); photoionization phenomena have not, therefore, generally been studied by the photochemist, since ionization energies often correspond to $\lambda < 105$ nm. The processes are, however, of the greatest importance in the upper atmosphere, where short-wavelength ultraviolet radiation from the Sun can lead to appreciable ionization of the chemical species present. There are, nevertheless, a few substances whose ionization potential is lower than the lithium fluoride cut-off, of which, perhaps, the most investigated is nitric oxide. At wavelengths shorter than 134.3 nm, absorption of light is followed by ionization:

$$NO + h\nu \rightarrow NO^+ + e \qquad (3.2)$$

Photoionization of some *excited* molecules, for which ionization of the ground state does not occur at $\lambda > 105$ nm, has also been observed, and it may be used to characterize and estimate the excited species. For example, electronically excited O_2 (in the $^1\Delta_g$ state) may be photoionized by argon resonance radiation, which is just transmitted by an LiF window.

3.3 Predissociation

Complex organic molecules do not usually undergo optical dissociation in the regions of strongest absorption. The increased number of electronic states, the closer spacing between them, and the large number of vibrational modes all tend to increase the probability of radiationless transitions between states. Thus, an excited state populated below its dissociation limit may undergo a radiationless transition to populate another state *above* its dissociation limit. The process is one that, for small molecules, is treated as predissociation, and in this section we describe the physical principles.

The word *predissociation* was adopted to describe the spectroscopic appearance of an absorption system. The gas-phase absorption spectra of the simpler molecules show considerable sharp rotational structure, but in some cases this rotational structure becomes blurred, leading to a diffuseness of the bands at a wavelength longer than that corresponding to the optical dissociation limit. For example, in the absorption spectrum of S_2 there is a region of diffuseness near the 17, 0 band, although the bands are sharp at other wavelengths.

Predissociation is now understood to arise from the 'crossing' of electronic states, and the occurrence of *radiationless* intramolecular energy transfer between them. Figure 3.3 shows the crossing curves for the S_2 molecule. The diffuseness arises from an increase in the line widths of the individual rotational transitions, and in order to explain it we must first consider the line width of a spectroscopic transition for which there is no radiationless transition. In the absence of molecular collisions, a species will remain in an excited state for an average time (the radiative lifetime, τ_0) of about $1/A$,

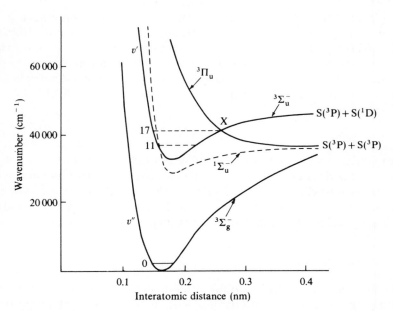

Fig. 3.3 Potential energy curves for S_2 showing the crossings that lead to 'normal' and 'induced' predissociations. (From Bowen, E. J. (1946). *Chemical aspects of light*, p. 95. Oxford University Press.)

where A is the Einstein factor for spontaneous emission. A spectral line has a minimum finite width—the *natural line width*—which is determined by the radiative lifetime through a mechanism related to the Heisenberg uncertainty principle. In terms of wavelength, the width $\Delta\lambda$ is given by

$$\Delta\lambda = \lambda^2/2\pi c\tau_0 \qquad (3.3)$$

For a permitted transition, $\tau_0 \sim 10^{-8}$ s, so that the natural line width is about 5×10^{-6} nm at $\lambda \sim 300$ nm, which is much less than the spacing between rotational lines. If, however, Δt becomes less than τ_0, the line width will increase. A decrease in lifetime of an excited state can be brought about by the occurrence of radiationless transitions to a different state. Radiationless transitions taking place during the period of a few vibrations reduce the effective lifetime to about 10^{-13} s, and $\Delta\lambda$ becomes about 0.5 nm, which is now *greater* than the rotational spacing: the spectrum appears diffuse.

The discussion of the last paragraph will have made it clear that for diffuseness of rotational fine structure to be observed, the radiationless transition must occur rapidly enough to give sufficient line-broadening. However, even if radiationless transition occurs only at ten times the rate of spontaneous emission (typically, $A < 10^8$ s^{-1}), most of the molecules in the first excited state will pass over to the second state, although there will be no apparent diffuseness in the absorption spectrum. Emission bands will, on the other hand, be drastically reduced in intensity under such circumstances, since most molecules will not survive in the first state long enough to radiate. Thus the 'breaking-off' of bands in emission is a more sensitive test of radiationless transition than is diffuseness in absorption: the term 'predissociation' refers to the transition processes leading to dissociation rather than to the diffuseness in absorption spectra.

Chemical dissociation of the absorber may follow radiationless transition if the transition occurs at an energy sufficient to cause dissociation of the new state. This energy may well be less than the dissociation energy in the state that is populated initially by absorption of radiation. For a diatomic molecule, the fragments from photodissociation must be chemically identical, whatever the dissociation mechanism. It follows that the state of excitation of fragments produced by predissociation below the optical dissociation limit must be lower than those formed by optical dissociation in a continuous absorption region. The solid potential energy curves for S_2 (Fig. 3.3) show that predissociation yields two ground-state atoms ($S(^3P)$), while optical dissociation would lead to one excited, $S(^1D)$, atom. It is important to note that, while predissociation may lead to the formation of some product at wavelengths longer than the dissociation limit, the products may well not be identical with those from optical dissociation.

The Franck–Condon principle (Section 2.7) applies to radiationless, as well as to radiative, transitions, as will be discussed in more detail in Section 4.5.

The nuclei do not move significantly during the course of the change in electronic state. If two potential curves intersect, then the two states possess the same total (electronic + vibrational) energy at the same internuclear distance. (If the curves do *not* intersect, either the internuclear distance must change (a violation of the Franck–Condon principle) or kinetic energy must be released instantaneously.) Radiationless transition occurs, therefore, at the internuclear distance and at the energy represented by the crossing point of two intersecting potential energy curves (e.g. the point X in Fig. 3.3). The actual rate of crossing is determined in part by whether the radiationless transition is 'allowed' or not, and in part by the overlap of the vibrational probability curves at and near the crossing point (cf. Section 2.7). Selection rules have been derived for radiationless transitions, analogous to those for optical transitions. They are, where applicable,

$$\Delta S = 0 \tag{3.4}$$

$$\Delta J = 0 \tag{3.5}$$

$$\Delta \Lambda = 0, \pm 1 \tag{3.6}$$

$$+ \leftrightarrow + \qquad g \leftrightarrow g$$
$$- \leftrightarrow - \qquad u \leftrightarrow u \tag{3.7}$$

Except for the rule $g \leftrightarrow g$, $u \leftrightarrow u$, the rules are identical with the optical transition selection rules (2.9), (2.12), and (2.13).

Another factor that determines the probability of crossing is the length of time spent by the molecule in the configuration of the crossing point. This duration is determined by the velocity of approach, which is a measure of the excess energy possessed by the molecule over that required for crossing. It is a frequent observation that a spectrum showing diffuseness as a result of predissociation becomes sharp again at shorter wavelengths.

3.4 Induced predissociation

The efficiency of crossing between two electronic states may be so low that predissociation no longer reduces the intensity of emission bands. Even in the absence of loss processes such as physical quenching, radiative loss will now ensure that most of the excited species do not undergo chemical change. Such inefficiency of intramolecular energy transfer between crossing states usually reflects some degree of 'forbiddenness' of the radiationless transition. There are, however, some cases where the efficiency of the radiationless transition depends on the external environment. Collisions with other species, or the presence of magnetic or electric fields, can invalidate the selection rules for *optical* transitions. Similar apparent breakdown of radiationless transition selection rules is observed—the rules are applicable only to unperturbed molecules. The increased probability of crossing between the appropriate states can lead to an increase in the relative extent of predissociation as a molecule becomes more perturbed by its external environment. Predissociation that is significant only in the presence of some perturbation is known as *induced predissociation*.

Collision-induced predissociation may be recognized by the broadening, caused by increased diffuseness, of bands in the absorption spectrum, seen when the pressure is increased or when a foreign gas is added to the absorbing system. Predissociations induced by the absorbing gas itself lead to deviations from the Beer–Lambert law: the apparent absorption increases faster than the law predicts. A similar apparent increase in absorption intensity is seen for predissociation induced by a foreign gas. (The sharper the lines in an absorption band, the weaker the band appears in a low-dispersion spectrum that does not resolve the individual lines, since the unabsorbed background contributes more to the total transmitted intensity.)

One of the first instances in which induced predissociation was recognized was in the I_2 molecule. We have already described the banded appearance of the absorption spectrum in terms of the $^3\Pi \leftarrow {}^1\Sigma$ transition (Fig. 3.1). At very low pressures, absorption in the banded region below the dissociation continuum is followed by emission of the fluorescence spectrum. There is, however, an electronic state to which radiationless transitions are nominally 'forbidden', which crosses the $^3\Pi$ state, and addition of a foreign gas (Ar at a pressure of, say, 30 mmHg) is found to weaken the emission bands above the predissociation limit. Indeed, atomic iodine—the product of the predissociation—may be detected (by the atomic absorption lines) under those conditions where the emission bands are quenched. An increase in apparent intensity of the appropriate absorption bands has also been observed in the presence of a foreign gas, and there is little doubt in this case that the effect is a result of the induced predissociation. The predissociation leads to the formation of two *ground*-state atoms, and can occur in the other halogens as well as in I_2 at wavelengths possessing enough energy to break the bond. Thus, from a study of the photobromination of ethene, the quantum yield for the primary step

$$\text{Br}_2 + h\nu \rightarrow 2\text{Br}(^2\text{P}_{1/2}) \tag{3.8}$$

has been shown to be unity for photolysis at wavelengths extending up to the energy of the dissociation limit ($190 \text{ kJ mol}^{-1} \equiv \lambda = 628.4 \text{ nm}$). In fact, the primary quantum yield is still unity at $\lambda = 680 \text{ nm}$, where the energy (176 kJ mol^{-1}) is insufficient to break the Br_2 bond. At this wavelength the extinction coefficient is temperature-dependent; the absorption appears to occur from levels with $v'' > 0$, and the photolysis affords an example of vibrational energy in the ground electronic state contributing to the energy required for dissociation.

One of the most dramatic examples of a predissociation induced by an external field, rather than by collision, is the quenching of iodine fluorescence in the presence of a magnetic field. The fluorescent emission in the visible spectrum is extinguished when a sufficiently strong magnetic field is applied. It has been shown that the selection rule $\Delta J = 0$ no longer holds strictly in the presence of a magnetic field, and crossing can take place to one of the predissociating states that correlate with two ground-state iodine atoms.

3.5   Radiationless transitions in complex molecules: intramolecular energy transfer (1)

Radiationless transitions are favoured in complex molecules for the reasons suggested at the beginning of Section 3.3, and processes involving such intramolecular energy transfer are the most likely route to photodissociation. Detailed discussion of energy transfer is deferred until Chapters 4 and 5, since both intramolecular and intermolecular energy transfer are studied for the most part in terms of the emission phenomena described in Chapter 4. The processes of predissociation and induced predissociation have, however, been illustrated with examples involving very simple molecules, and we must see how far the photochemical dissociation of more complex species follows similar mechanisms.

Two main obstacles prevent description of the photochemistry of large molecules in such precise physical terms as those possible for simple molecules. First, the absorption spectrum of a complex species may not show

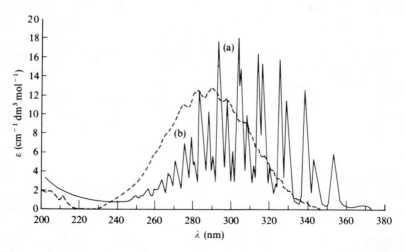

Fig. 3.4   Absorption spectra of gaseous formaldehyde (a) and acetaldehyde (b), showing the loss of resolvable structure with increasing molecular complexity. (From Calvert, J. G. and Pitts, J. N., Jr (1966). *Photochemistry*, p. 368. Wiley, New York.)

sufficient resolvable structure either for an identification of state to be possible or for optical dissociation and predissociation to be recognized. The relative lack of spectroscopic structure is, of course, a result both of the greater complexity and closer spacing of vibrational and rotational levels, and of the increased number of electronic states. Figure 3.4 illustrates the loss of resolvable structure in going from formaldehyde to acetaldehyde. Secondly, a considerable number of fragmentation paths may exist for an excited polyatomic molecule. Although the relative importance of each path may depend on the wavelength of the dissociating radiation, several different sets of primary products are often produced simultaneously. In addition, reorganization of the internal energy of an excited molecule can lead to rupture of a bond remote from the part of the molecule where the excitation was first localized. (In long-chain aldehydes, for example, cleavage can occur 'down-chain' from the carbonyl group that absorbs the energy: the cleavage may not even result directly from an electronically excited state, but rather from a high level of vibrational excitation in the ground electronic state.) To illustrate the occurrence of several photodissociative steps, Table 3.1 gives the approximate quantum yields, where known, for five *primary* processes in the photolysis of butanal at different wavelengths (data taken from Calvert J. G. and Pitts, J. N., Jr (1966). *Photochemistry*, pp. 372–3, Wiley, New York).

Table 3.1 *Approximate quantum yields for primary steps in photolysis of butanal*

Products	Quantum yield				
λ (nm) =	313.0	280.4	265.4	253.7	187.0
n-C_3H_7 + HCO	$\geqslant$0.35	$\geqslant$0.28	$\geqslant$0.28	0.31	0.34
C_3H_8 + CO	0.017	$\geqslant$0.11	$\geqslant$0.25	0.33	0.13
C_2H_4 + CH_3CHO	0.16	0.27	0.38	0.30	0.24
CH_3 + CH_2CH_2CHO	0.005	0.006	0.110	0.015	0.25
C_3H_6 + CH_2O	—	—	—	—	0.04

Although it may prove impossible to make direct spectroscopic assignments to the transition in an absorption, there are several ways in which the general type of the transition can be adduced. The maximum value of the measured extinction coefficient, ε_{max}, may be used to determine the degree of 'forbiddenness' in a transition. In Section 2.4 it was suggested that $\varepsilon_{max} \sim 10^5 \text{ cm}^{-1} \text{ dm}^3 \text{ mol}^{-1}$ for a totally permitted transition (this value is dependent, of course, on the width of the absorption band, and is only an order-of-magnitude number). Thus the near-ultraviolet absorption bands ($\lambda \sim 240 - 340$ nm) of the aldehydes (cf. Fig. 3.4) have values of ε_{max} in the range 12–20 (all units are for molar decadic extinction coefficients), and are presumed to have their origins in a symmetry-forbidden transition (a singlet→singlet n→π* transition). Another absorption band at $\lambda \sim 400$ nm is extremely weak ($\varepsilon_{max} \sim 10^{-3}$), and is probably due to a transition that is both spin- and symmetry-forbidden. Most organic molecules are singlets in their ground states, and very small values of ε_{max} may well suggest that the electronic states produced on absorption are triplets; evidence from emission processes can often confirm such a hypothesis. Table 3.2 gives a rough idea of the reduction in extinction coefficient resulting from a contravention of certain selection rules; the factors are, of course, multiplicative if several selection rules are broken simultaneously.

Table 3.2 *Extinction coefficients for various types of transition (From Platt, J. R. (1953) J. Opt. Soc. Am. 43, 252)*

Type of transition	Decadic molar extinction coefficient ε (cm^{-1} dm^3 mol^{-1})	Specific system
Allowed	10^5	$\pi \to \pi*$
Spin-forbidden ($\Delta S \neq 0$)	1	Second row elements
Overlap-forbidden (electronic charge changes position)	10^3	n → π* in second row heteroatoms
Momentum-forbidden (large changes in linear or angular momentum)	10^4–10^2	Condensed ring systems
Parity-forbidden (g↔u)	10^4	Condensed ring systems

In molecules for which both n → π* and π→π* transitions are possible the two possibilities may often be distinguished by the effect that polar and non-polar solvents have on the wavelengths of the absorption band. Only one n electron is available in an excited n, π* state, and polar solvents are more strongly hydrogen-bonded to the ground state than to the excited state. The energy of the ground state is therefore lowered more by the solvent interaction than is the excited state, the energy gap between the two levels is increased, and a shift of the absorption spectrum to shorter wavelengths ('blue shift') is observed in polar solvents. A *red* shift is observed in some π → π* transitions (e.g. in ketones) in polar solvents; the excited state is *more* polar than the ground state, and interaction with a polar solvent reduces the energy difference between the two states (cf. Section 6.2).

If it is possible to produce a single crystal, or an isotropic distribution of molecules in a rigid glass, of a molecule with a plane of symmetry, so that the orientation of the crystal plane to the molecular axes is known, then the interaction with polarized light can reveal the nature of the spectroscopic transition. For example, the $\pi \to \pi^*$ transition in pyridine is polarized in the plane of the molecule, while the $n \to \pi^*$ transition is polarized perpendicular to it. Kasha has suggested a list of criteria for assignment of a transition as $n \to \pi^*$ rather than as $\pi \to \pi^*$. These criteria are: (a) an $n \to \pi^*$ transition is absent in analogous hydrocarbons; (b) $n \to \pi^*$ bands disappear in acid media; (c) $n \to \pi^*$ bands show a 'blue' shift in polar solvents; (d) conjugative substituents also cause a blue shift; (e) $n \to \pi^*$ bands are found at relatively long wavelengths compared with $\pi \to \pi^*$ bands; (f) $n \to \pi^*$ bands are relatively weak compared with $\pi \to \pi^*$ bands; and (g) the transitions may show specific polarization (e.g. the $n \to \pi^*$ transition in pyridine is polarized perpendicular to the plane of the ring).

The preceding discussion will have made it clear that potential energy curves (or, rather, sections through the many-dimensioned potential energy hypersurface) cannot usually be constructed for complex polyatomic molecules. Figure 3.5 shows an alternative diagram for a complex molecule such as naphthalene. The diagram is a modification of a form devised by Jablonski, and is frequently called a *Jablonski diagram*. It does not attempt to represent the molecular shapes and sizes, and the vibrational levels drawn for each state do not usually correspond to the actual v'', v' numberings and spacings. On the other hand, the energies of the vibrational ground states of each electronic level are shown correctly if the experimental evidence is available. As will be seen, the $S_0, S_1, \ldots, T_1, \ldots$ notation is employed. Wavy lines on the diagram represent radiationless energy conversion: the vertical wavy lines within a particular electronic state indicate degradation of vibrational excitation (probably by a collisional, *inter*molecular process), while the horizontal wavy lines indicate intramolecular energy exchange. Formal distinction is drawn between electronic energy exchange permitted by the $\Delta S = 0$ rule and that forbidden by it. The term *internal conversion* (IC) is applied to radiationless transitions between states of the same spin multiplicity, while *intersystem crossing* (ISC) refers to energy exchange between states belonging to different (spin) 'systems'. Both internal conversion and intersystem crossing are assumed to take place with no change in total electronic + vibrational energy, and the wavy lines are therefore horizontal (i.e. no translational or rotational energy is released in an intramolecular electronic energy exchange—see Section 4.5).

Jablonski diagrams are, in fact, more frequently used in connection with discussions of luminescence, but are introduced in this chapter to illustrate the similarity between predissociative fragmentation in simple molecules and the route to photodissociation in complex ones. The process of internal conversion from $S_1(v=0)$ to S_0 shown in Fig. 3.5 leads to a high degree of vibrational excitation in the ground electronic state. If sufficient energy is available, the molecule may then undergo spontaneous dissociation (or isomerization, etc.). The condition of a molecule in our example (where S_0^v is

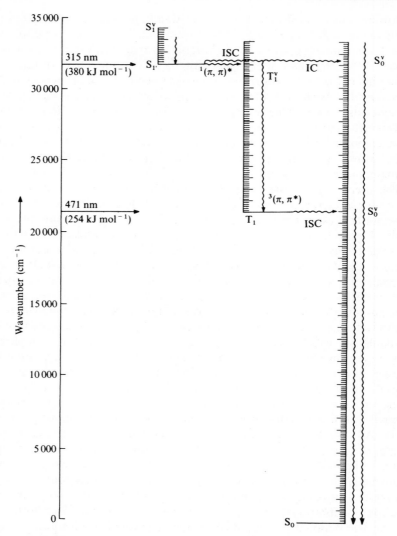

Fig. 3.5 Method of representing energy levels and electronic states for a complex molecule such as naphthalene. This is a modified 'Jablonski diagram'.

populated) is identical to the condition of a molecule thermally 'activated' by collision. If enough energy to break a bond is stored somewhere in the molecule, then, so long as molecular collision does not *deactivate* the molecule, dissociation may occur when the energy has accumulated in the bond. The competition between reaction and deactivation, and the way in which energy accumulates in a bond, is treated by the several theories of unimolecular reaction (e.g. 'free-flow' or 'strictly harmonic'), and further discussion is out of place here. Suffice it to say that photochemical activation is an almost ideal technique for the production of virtually monoenergetic species with which the theories may be tested. Several examples are known in which the rate of unimolecular decomposition shows an increase with decrease in wavelength of photolysis that is exactly predictable on a theoretical basis.

The essential differences between predissociation of a diatomic molecule and the mechanism we have just outlined for photodissociation of a complex entity should now become clear. If some state of a diatomic molecule is populated, by intramolecular energy transfer, with enough energy to dissociate, then it is likely to dissociate in the first vibration (i.e. in about 10^{-13} s); if it is populated at a level below the dissociation limit of this state, then (except under special circumstances) it will not dissociate. In a large molecule it is necessary to make three major modifications to this description. First, any one of a number of bonds may be broken, and, further, intramolecular elimination reactions can occur. Secondly, the time taken for the energy to flow to the bond(s) involved may be orders of magnitude greater than the 10^{-13} s taken for the diatomic molecule to dissociate, and the probability that the excited molecule may be deactivated before it can dissociate is increased accordingly. Thirdly, thermal energy already possessed in vibration of other bonds of the molecule can contribute to the total energy required for disruption of the reactive bond.

3.6 Important primary dissociation processes

Dissociation of a diatomic molecule can only yield two atomic fragments whose chemical identity is not in doubt. However, the dissociation of a polyatomic molecule can sometimes give many sets of products, so that an understanding of the photochemistry of large molecules requires that the nature of the primary products be known. Direct identification is sometimes possible: alternatively, the chemistry of the primary step may have to be inferred from the final products of reaction. For a more detailed discussion of the methods used to elucidate primary dissociative mechanisms, the reader is referred first to the books of Calvert and Pitts, and of Okabe (see Bibliography), and then to the original references cited. We indicate the types of process that occur with small inorganic species in Table 3.3. Photodissociative processes in organic species are illustrated with reference to two classes of compound: hydrocarbons and carbonyl compounds.

3.6.1 *Hydrocarbons*

The alkanes absorb strongly in the 'vacuum' ultraviolet region: methane starts to absorb around 144 nm, and the higher alkanes absorb at

Table 3.3 Primary dissociative processes in the photochemistry of some inorganic molecules. (Data obtained from Calvert J. G. and Pitts, J. N., Jr (1966). Photochemistry. Wiley, New York, in which original references may be found)

Species	Products[a]		Wavelength[b] λ (nm)	Quantum yield[c]	Remarks
Hydrides					
H_2O	$H + OH(^2\Pi)$	(1)	<242	~1	$OH(^2\Sigma \rightarrow {}^2\Pi)$ fluorescence observed.
	$H + OH(^2\Sigma)$	(2)	<135.6	—	
	$H_2 + O(^1D)$	(3)	123.6	$\phi/(\phi_1 + \phi_2) \sim 0.3$	
H_2O_2	$2OH(^2\Pi)$	(4)	253.7	0.85 ± 0.2	
	$OH(^2\Pi) + OH(^2\Sigma)$	(5)	<202.5	—	$OH(^2\Sigma \rightarrow {}^2\Pi)$ fluorescence observed.
H_2S	$H + SH$	(6)	200–255	~1	
NH_3	$NH_2 + H$	(7)	<217	96% of reaction at $\lambda = 184.9$ nm	Total ϕ for NH_3 disappearance ~1 at $\lambda = 184.9$ nm and at low pressures. At $\lambda = 147$ nm, and at $p = 15$ mmHg, $\phi = 0.45 \pm 0.1$.
	$NH(^3\Sigma^-) + 2H$	(8)	<155		
	$NH(^1\Pi) + H_2$	(9)	<129.5	14% of reaction at $\lambda = 123.6$ nm	
	$NH_3^+ + e^-$	(10)	<123		
HN_3	$N_2 + NH$	(11)	~200		
$NH_2 \cdot NH_2$	$NH_2 + NH_2$	(12)	199	~1	Possible alternative split to $H + NHNH_2$.
HI	$H + I$	(13)	<327	~1	Fraction of I in $^2P_{1/2}$ and $^2P_{3/2}$ states at various wavelengths not known. Similar processes for HBr, HCl.

Oxides	Products		ϕ	λ (nm)	Comments
O_2	$O(^3P)+O(^3P)$	(14)		~ 245.4	Via forbidden absorption to the $^3\Sigma_u^+$ state of O_2.
	$O(^3P)+O(^1D)$	(15)	~ 1	< 175.9	Via allowed absorption to $^3\Sigma_u^-$ in Schumann–Runge continuum.
	$O(^3P)+O(^1S)$	(16)		< 134.2	Suggested from appearance of spectrum. $2O(^1S)$ can be formed at $\lambda < 92.3$ nm.
O_3	$O(^1D)+O_2(^1\Delta_g)$	(17)	~ 1	< 310	
	$O(^1D)+O_2(^1\Sigma_g^+)$	(18)	—	< 266	
	$O(^3P)+O_2(^3\Sigma_g^-)$	(19)	~ 1	~ 600	
	$SO+O$	(20)	—	< 218	
SO_2	$SO_2+O(^3P)$	(21)		< 344	At $\lambda < 224$ nm, $O(^1D)$ can be formed.
SO_3	$SO+O_2(^3\Sigma_g^-)$	(22)		< 300	SO observed in flash photolysis but could be product of secondary reaction.
N_2O	$N_2(^1\Sigma_g^+)+O(^1D)$	(23)	~ 1	~ 180	
	$N+NO$	(24)	12% at $\lambda = 123.6$ nm		
NO	$N(^4S)+O(^3P)$	(25)		183.2	
	$N(^2D)+O(^3P)$	(26)		123–140	
	$N(^2P)+O(^3P)$	(27)		< 123	
	$NO^+ + e^-$	(28)		< 134.3	

(Table continued on next page)

Table 3.3 (*continued*)

Species	Products[a]		Wavelength[b] λ (nm)	Quantum yield[c]	Remarks
NO_2	$NO + O(^3P)$	(29)	<400		
	$NO + O(^1D)$	(30)	228.8		
NOCl	$NO + Cl$	(31)	<760	~ 1 at $\lambda = 253.7$ nm	
CO_2	$CO(^1\Sigma^+) + O(^1D)$	(32)	<165	~ 1 for $\lambda = 123.6\,-$	
	$CO(^1\Sigma^+) + O(^1S)$	(33)	<127.3	150 nm	
	$CO(^3\Pi) + O(^3P)$	(34)	<107	$= 1.2 \pm 0.1$ at 123.6 nm	
C_3O_2	$CO + C_2O$	(35)	~ 300	—	
Cl_2O	$ClO + Cl$	(36)	220–850	—	
ClO_2	$ClO + O(^3P)$	(37)	~ 375.3	—	
	$ClO + O(^1D)$	(38)	—		
Halogens					
I_2	$I(^2P_{3/2}) + I(^2P_{1/2})$	(39)	<499	~ 1	Optical dissociation.
	$2I(^3P_{3/2})$	(40)	<803.7		Induced predissociation.
					Similar processes for other halogens.

[a] The term symbol is not normally given if a species is formed in its ground electronic state; if, however, both ground state and electronically excited states of the same product can appear, then the states are all specified.

[b] The wavelengths normally refer to those for which a process is energetically possible. If, however, a quantum yield is quoted, the wavelengths refer to those at which the quantum yield was measured or, alternatively, wavelength is specified in the 'quantum yield' column.

[c] Or percentage of total reaction proceeding to the given products.

progressively longer wavelengths (e.g. n-butane shows absorption at $\lambda < 166$ nm). The maximum decadic molar extinction coefficients are about 10^4, and the absorption is thought to be due to an allowed $\sigma \rightarrow \sigma*$ transition. In the wavelength region 129.5–147 nm, molecular elimination of hydrogen is the most important photodissociative process,

$$RCH_2R' + h\nu \rightarrow RCR' + H_2 \tag{3.9}$$

although bond ruptures at almost any point are minor processes and can lead to the formation of hydrogen atoms and a variety of free radicals. For CH_4, the elimination process (3.9) is nearly six times more frequent at $\lambda = 123.6$ nm than the radical fission

$$CH_4 + h\nu \rightarrow CH_3 + H \tag{3.10}$$

Photoionization occurs at shorter wavelengths (for CH_4 with $\lambda < 96.7$ nm).

The lowest-energy singlet $\rightarrow$ singlet ($\pi \rightarrow \pi*$) absorption band is found at longer wavelengths in unsaturated hydrocarbons than in the paraffins. For ethene and substituted ethenes the absorption maximum lies around 180 nm, while conjugation shifts the spectrum even further towards the visible: the compound $CH_3(CH=CH)_{10}CH_3$ has an absorption maximum at $\lambda \sim 476$ nm. Isomerization (Chapter 6) is a frequent fate of the excited state formed on absorption, but fragmentation is also observed. For example, in ethene, the processes

$$CH_2 = CH_2 + h\nu \rightarrow H_2 + H_2C = C\text{:}(\rightarrow HC \equiv CH) \tag{3.11}$$

$$\rightarrow 2H + H_2C = C\text{:} \tag{3.12}$$

$$\rightarrow H_2 + HC \equiv CH \tag{3.13}$$

$$\rightarrow 2H + HC \equiv CH \tag{3.14}$$

are thought to take place over the wavelength range 123.6–184.9 nm. Ethyne is formed in reaction (3.11), via $H_2C = C\text{:}$, about 1.5 times more rapidly than in the direct process (3.13) at $\lambda = 147$ nm. At the same wavelength the radical (3.12, 3.14) and molecular (3.11, 3.13) mechanisms occur with almost equal efficiency; fission to $CH_2CH + H$ is much less frequent.

Fragmentation of the polyenes is important in low-pressure gas-phase photolysis, but is suppressed in the presence of added gas. It would appear that bond rupture occurs in a vibrationally excited ground-state molecule that is formed by intersystem crossing from the excited singlet state.

Simple aromatic hydrocarbons possess a moderately strong absorption in the near ultraviolet: for benzene the absorption is of the symmetry-forbidden $^1B_{2u} \leftarrow {}^1A_{1g}$ system and has $\varepsilon_{max} \sim 160$ at $\lambda_{max} = 256$ nm. An intense allowed absorption band is observed at shorter wavelengths (in benzene the transition of the $^1E_{2u} \leftarrow {}^1A_{1g}$ system lies at $\lambda \sim 180$ nm). Absorption in the longer-wavelength band leads mainly to emission of radiation (Chapter 4) or to reaction of the excited species (Chapter 6), and quantum yields for photodissociation are very small. However, in the gas phase at $\lambda = 184.9$ nm the quantum yield for benzene disappearance is near unity, and polymer, carbon, and traces of volatile products are formed.

3.6.2 *Compounds containing the carbonyl group*

Photochemical data are probably more extensive for compounds containing the carbonyl, C=O, group than for any other class of organic compound. In the interest of brevity, we shall confine our remarks here mainly to the photochemistry of aldehydes and ketones, since, broadly speaking, the acids, acid anhydrides, esters, and even amides undergo analogous photodissociative reactions.

In each class of carbonyl-containing compound, the first absorption is of the forbidden $n \rightarrow \pi^*$ transition. In aliphatic aldehydes the maximum of this band·lies near 290 nm (cf. Fig. 3.4), while in ketones it is displaced to slightly shorter wavelengths (~ 280 nm); aromatic substitution shifts the absorption to longer wavelengths (e.g. $\lambda_{max} \sim 340$ nm in benzophenone). The absorption lies at considerably shorter wavelengths in acids, anhydrides and esters (< 250 nm), and in amides (< 260 nm). Allowed $\pi \rightarrow \pi^*$ and $n \rightarrow \sigma^*$ transitions give rise to intense absorption bands at short wavelengths (e.g. around 180 and 160 nm for aldehydes).

R. G. W. Norrish and co-workers were the first to make a systematic study of the photochemistry of aldehydes and ketones, and it is for this reason that certain possible reaction paths are commonly known as *Norrish Type I*, *Norrish Type II*, etc. (the general formulation was first given by Norrish and Bamford (1937), *Nature*, **140**, 195).

The important *dissociative* primary photochemical processes in ketones can be represented as

$$RCOR' + h\nu \rightarrow R + (COR')^*$$
$$\longrightarrow R' + CO \qquad (3.15a)$$

$$\rightarrow (RCO)^* + R'$$
$$\longrightarrow R + CO \qquad (3.15b)$$

$$R_2CHCR_2CR_2COR' \rightarrow R_2C=CR_2 + CR_2 = C(OH)R' \qquad (3.16)$$
$$\longrightarrow CHR_2COR'$$

The radical-forming processes (3.15) are those known as Norrish Type I, while the intramolecular fission of the C–C bond α–β to the carbonyl group is the Norrish Type II process.

The relative importance of the two paths (3.15a) and (3.15b) in Type I photolysis seems to depend on the wavelength of photolysis. Thus, in the photolysis of butan-2-one,

$$CH_3COC_2H_5 + h\nu \xrightarrow{\ a\ } CH_3CO + C_2H_5 \qquad (3.17a)$$

$$\xrightarrow{\ b\ } CH_3 + COC_2H_5 \qquad (3.17b)$$

the relative contribution of paths a and b varies from 40:1 at $\lambda = 313$ nm to 2.4:1 at $\lambda = 253.7$ nm.

Some thermal equilibrium of hot acyl radicals may occur, but almost immediate decomposition to alkyl radical and carbon monoxide becomes increasingly likely as the wavelength of photolysis is decreased.

Type II photolysis in ketones has been shown fairly conclusively to proceed via a cyclic six-membered intermediate. For example in pentan-2-one the process may be represented by the equation

$$(3.18)$$

The *enol* thus formed then tautomerizes to propanone.

Aldehydes photodissociate according to a 'Type I', and where appropriate a 'Type II', mechanism. An intramolecular elimination reaction, not observed with ketones, also occurs:

$$RCHO + h\nu \rightarrow RH + CO \qquad (3.19)$$

For the simple aliphatic aldehydes, at a fixed wavelength the quantum yield of reaction (3.19) bears a constant ratio to the quantum yield for 'Type I' fission. The ratio is, however, wavelength-dependent, being less than 0.05 at $\lambda = 313$ nm, 0.34 at $\lambda = 280.4$ nm, 0.95 at $\lambda = 265.4$ nm, and 1.22 at $\lambda = 253.7$ nm.

Cyclic ketones decompose under the influence of light in a manner which suggests that a biradical is formed in the primary step. Cyclopentanone, for example, is photodissociated in the gas phase in three ways:

$$2C_2H_4 + CO \qquad (3.20)$$

$$\longrightarrow \quad \square \quad + \quad CO \qquad (3.21)$$

$$\longrightarrow \qquad CH_2{=}CHCH_2CH_2CHO \qquad (3.22)$$

The primary products may certainly be interpreted in terms of initial form-ation of $\cdot CH_2CH_2CH_2CH_2CO\cdot$. It appears that ethene is formed directly from $\cdot CH_2CH_2CH_2CH_2\cdot$ (derived from the primary biradical by loss of CO) rather than from decomposition of 'hot' cyclobutane after ring closure. There is, however, considerable evidence that vibrationally excited molecules are of importance in the gas-phase photolysis.

3.7   Photochemistry in solution

Much of what has been said about photochemical processes so far has referred specifically to gas-phase photochemistry. This seems to be an appro-priate point at which to indicate the differences between processes in the gas phase and those in condensed phases (particularly in solution).

Both the act of absorption and the subsequent fate of excitation may be profoundly affected by going from the gaseous to the liquid state. We shall deal with the spectroscopy and the primary processes in turn.

The influence of polar and non-polar solvents on different types of elec-tronic transition has already been mentioned in Section 3.5. Molecular energies may be reduced by solvation, and the extent of this reduction depends both on the nature of the electronic state and on the nature of the liquid phase or solvent. Absorption intensities, especially of nominally forbid-den bands, may be affected by perturbations brought about by the proximity of solvent molecules. Collision broadening and the absence of well-defined rotational energy levels in the liquid phase may make it difficult to demon-strate the existence of a predissociation, even of a simple molecule, by other than circumstantial evidence. Such evidence sometimes suggests that the wavelength of the onset of predissociation may be different in the gaseous and liquid states. Thus, in going from gas phase to solution, the wavelength at which a photochemical process occurs, and the importance of that process relative to other concurrent processes, may be changed.

Several changes are also apparent in the fates of the electronically excited species formed on absorption. First, the large rate of molecular collision in condensed phases makes the probability of physical quenching of electron-ically (or vibrationally) excited species much higher than in the gas phase, and primary quantum efficiencies may be correspondingly small in solution. Secondly, *chemical* quenching of excited species may occur, i.e. the excited molecule may be removed by chemical reaction. In solution, the high concen-tration of solvent molecules may make this process competitive with fragmentation, or indeed any other loss process. (The reactions of excited species are discussed in Chapter 6.) Thirdly, the fragments formed in a dissociative process may themselves react with solvent every time they appear, and the *overall* step may then be regarded for practical purposes as the primary one. This kind of process is particularly important for free radicals produced in a hydrogen-containing solvent: hydrogenation of the radical effectively prevents the *free* radical from appearing (although, of course, a solvent radical is usually created in the reaction). Thus, the photolysis of $(CH_3)_2CH \cdot CO \cdot CH(CH_3)_2$ in solution yields CO, C_3H_8, and $(CH_3)_2CHCHO$ as the main Type I products. The overall efficiency of the process is low, for the reasons to be discussed in the next paragraph, although the quantum yield for the Type I decomposition increases with temperature and reaches about 0.3 at 96 °C. No C_6H_{14} or $C_3H_7CO \cdot COC_3H_7$ products appear in the solution-phase photolysis, the C_3H_7 and C_3H_7CO fragments yielding products only by hydrogen atom abstraction.

A very important factor in the differing photochemical behaviour in the gas phase and in solution is the occurrence of *cage effects* in the latter phase. In the liquid phase, collisions between species are not uniformly distributed in time, but occur in sets or 'encounters'. The colliding partners are enclosed in a solvent 'cage', which tends to prevent their separation by diffusion, and they may make several mutual collisions before leaving each other's sphere of influence. The rates of chemical reactions may be affected considerably by the occurrence of many collisions in an encounter. In particular, radical recombination may occur at every encounter since excess vibrational energy in a newly formed bond can be removed by collision with solvent molecules. 'Primary' recombination of radicals, formed by dissociation of an electronically excited molecule, is therefore of frequent occurrence. It occurs within 10^{-11} s of the generation of the radicals, before they have separated by a molecular diameter. This so-called *geminate recombination* of radicals within a solvent cage can greatly reduce the effective quantum yield for the dissociation process, since the reactant may be regenerated by primary radical recombination much more often than the radicals can escape from the solvent cage. The importance of the effect is related to the kinetic energy of the fragments formed on photolysis, as well as to the viscosity of the solvent. If the fragments have sufficient energy, they may force their way out of the cage, and the primary quantum yield of some solution-phase photolyses increases as the wavelength of photolysing radiation is decreased.

Cage effects are responsible for the interesting phenomenon of chemically induced dynamic polarization observed in photochemical magnetic resonance experiments. Chemically induced dynamic nuclear spin polarization (CIDNP) manifests itself by abnormal absorption intensities, and even emissions, in NMR studies of photochemical processes. In generating the CIDNP phenomenon, two recombining radicals formed photolytically yield a product that initially possesses a non-equilibrium distribution of nuclear spin states, so that the NMR signals may be of unusual intensity. The ordinary spectrum reappears within seconds of shutting off the photolytic radiation. Because the electron spins of the recombining radicals must pair to give the (singlet) ground-state product, spin polarization accompanies the interaction within the solvent cage and is coupled to the nuclear spin states, thus giving rise to the CIDNP effect. Products formed within the solvent cage and those formed after radical separation are distinguishable (as may also be the spin multiplicities of the initial radical pair). The CIDNP technique thus affords one method for the diagnostic study of certain types of photochemical reaction. A related dynamic polarization phenomenon is worthy of note. Chemically induced dynamic electron spin polarization (CIDEP) probably arises as a result of different intersystem crossing rates from the state first formed by absorption to the sublevels of a triplet state. Electron spin resonance (ESR) emission may be observed, and the phenomenon can yield information about the energetics and relaxation of triplet states.

3.8 Photochemistry of ionic species

A further distinctive feature of solution-phase photochemistry, especially where the solvent is water, is the importance of ionic species. 'Primary' photochemical reaction involving ions in solution is somewhat different from the photodissociations of neutral molecules, since the process often involves oxidation–reduction steps.

The colours of transition metal ions are in general the result of symmetry-forbidden optical transitions of electrons on the metal atom; the forbiddenness is revealed by relatively weak absorptions (e.g. the absorption that causes the blue colour of the free Cu^{2+} ion has $\varepsilon_{max} \sim 10$ at $\lambda = 810$ nm). Many ions do, however, have an *intense* absorption ($\varepsilon_{max} \sim 10^4$) in the ultraviolet region (usually, but not always, in the wavelength range 200–250 nm). It is fairly clear that these absorptions derive from *charge-transfer* transitions, i.e. from transitions in which an electron is transferred either from one ion to another or from an ion to the solvent. The dark colour of the complex halides of Cu^{2+}, as distinct from the light blue of hydrated Cu^{2+}, is due to the tail of such a charge-transfer spectrum, which lies mainly in the ultraviolet.

Ionizing radiation can strip an electron from water (the process is *radiolysis*), and a comparison of the reducing species formed by radiolysis with that produced by the photolysis of aqueous solutions of ions has suggested that the species is, in fact, a hydrated electron. A transient absorption at around 700 nm is observed on flash photolysis (see Chapter 7) of aqueous ionic solutions, and it appears identical with the absorption seen on pulse radiolysis of pure water. The rates of reaction of the species produced in the two ways are also often identical. Furthermore, optical and electron paramagnetic resonance spectra of UV-irradiated aqueous glasses containing ionic species, and of trapped electrons produced by ionizing radiation, are identical. It appears, therefore, that hydration of the electron can make photodetachment of an electron energetically favourable at wavelengths much longer than those needed to cause photoionization in the gas phase. Quantum yields for the formation of hydrated electrons have been estimated, and may be relatively high. For photolysis of the halide ions (I^-, $\lambda = 253.7$ nm; Br^-, Cl^-, $\lambda = 184.9$ nm), for example, the quantum yields are probably 0.3–0.5. The quantum yield for the process

$$Fe(CN)_6^{4-} + hv \rightarrow Fe(CN)_6^{3-} + e_{aq} \tag{3.23}$$

may even approach unity ($\lambda = 200\text{--}260$ nm), although it is interesting that the quantum yield ($\lambda = 253.7$ nm) for

$$Fe^{2+} + hv \rightarrow Fe^{3+} + e_{aq} \tag{3.24}$$

is less than 0.1.

If the hydrated electron is involved, then the photochemical process may be written

$$X^{n+} + hv \xrightarrow{H_2O} X^{(n+1)+} + H_2O^- \tag{3.25}$$

The hydrated electron may then spontaneously dissociate,

$$H_2O^- \rightleftharpoons OH^- + H \tag{3.26}$$

or, more particularly, in acid solution it may react according to

$$H_2O^- + H^+ \rightarrow H_2O + H \tag{3.27}$$

In either case, atomic hydrogen is produced, and it can initiate secondary radical reactions. The photochemical decomposition of the I^- ion, for example, shows a dependence on pH that is consistent with a chain mechanism initiated by

$$I^- + hv \xrightarrow{H_2O} I + H_2O^- \rightarrow I + H + OH^- \tag{3.28}$$

and the low *overall* efficiency can be attributed to recombination in the cage of H and I atoms.

Charge transfer in cations can occur either to or from the solvent. Thus, absorption bands for both types of charge transfer[†] are known in iron-containing solutions:

$$Fe^{2+} \cdot H_2O + hv \rightarrow Fe^{3+} \cdot H_2O^-, \quad \lambda_{max} \sim 285 \text{ nm} \tag{3.29}$$

$$Fe^{3+} \cdot H_2O + hv \rightarrow Fe^{2+} \cdot H_2O^+, \quad \lambda_{max} \sim 230 \text{ nm} \tag{3.30}$$

There is some evidence to suggest that the longest wavelength absorption usually corresponds to the direction of the oxidation–reduction reaction that

---

[†] Although equations showing transfer of charge are *written* as if an electron is completely transferred, it must be understood that they may also represent partial electron transfer.

proceeds most easily. The reactions of H_2O^+, analogous to (3.26) and (3.27), are suggested to be

$$H_2O^+ \rightleftharpoons OH + H^+ \tag{3.31}$$

and

$$H_2O^+ + OH^- \rightarrow OH + H_2O \tag{3.32}$$

In systems containing ion-pair complexes, the energetics may be favourable for charge transfer from one partner (usually the cationic atom) to the other. The absorption maxima for the ion-pair charge-transfer spectra generally lie at longer wavelengths than the maxima for the uncomplexed cations (e.g. for the complex $Fe^{3+}CNS^-$, $\lambda_{max} \sim 460$ nm, compared with $\lambda_{max} \sim 230$ nm for charge transfer to Fe^{3+} from water). Typical processes involving Fe^{3+} complexes are:

$$Fe^{3+}CNS^- + h\nu \rightarrow [Fe^{2+}CNS] \rightarrow Fe^{2+} + CNS \tag{3.33}$$

$$Fe^{3+}Cl^- + h\nu \rightarrow [Fe^{2+}Cl] \rightarrow Fe^{2+} + Cl \tag{3.34}$$

$$Fe^{3+}OH^- + h\nu \rightarrow [Fe^{2+}OH] \rightarrow Fe^{2+} + OH \tag{3.35}$$

Atom production in reaction (3.34) has been demonstrated by the photo-initiation of vinyl polymerization by a radical mechanism in an iron (III) chloride system; it has also been shown that photochemical oxidations may be initiated by reaction (3.35) followed by hydrogen abstraction

$$OH + RH \rightarrow H_2O + R \tag{3.36}$$

Charge transfer occurs in two substances, $UO_2C_2O_4$ and $K_3Fe(C_2O_4)_3$, which are frequently used as *chemical actinometers* (see Chapter 7). In $UO_2C_2O_4$, charge transfer from UO_2^{2+} to $C_2O_4^{2-}$ ions leads to decomposition of the $C_2O_4^{2-}$; in $K_3Fe(C_2O_4)_3$, the change of importance is the reduction of Fe^{3+} ions to Fe^{2+}.

3.9   Multiphoton dissociation and ionization

The idea of multiphoton processes at first sight seems to strike at the foundations of quantum theory. Einstein showed that the photoelectric effect was consistent with the radiation behaving as quantized photons whose energy was determined by their frequency or wavelength; the intensity of radiation described the number of photons (per unit time), but did *not* affect the energy of any individual photon. Similar considerations applied to photochemical change; the Stark–Einstein law presented in Section 1.2 was taken by its proponents to provide further evidence for the ideas of quantization. One, and only one, photon was to be absorbed by a particle to produce photochemical change. As a consequence, photons of energy less than that needed to cause a particular change, such as dissociation, could not be effective, however high their intensity. Perhaps even more obviously, if the frequency of the radiation did not correspond to the difference between two energy levels in the molecule or atom, then absorption, and thus reaction, could not occur. Fairly recently, however, a large number of experiments have been performed in which *more than one* photon is absorbed by a single particle. Multiphoton excitation is a *non-linear* process, observed only with high intensities of light. It is the advent of lasers (cf. Sections 5.7 and 7.2) that has made the multiphoton experiments possible. Of course, the laws of quantization are not really being broken, but the simple view has to be modified to cope with the situation that arises when photons can arrive in very rapid succession at any given particle.

Two different kinds of mechanism can be envisaged for multiphoton excitation. We shall illustrate these mechanisms first for two-photon excitation, and it will be apparent that the ideas developed for the biphotonic process can readily be extended to the more general multiphoton case. *Sequential* excitation involves a real intermediate state of the absorbing species. That state can become populated by the first photon, and it can act as the starting point for the absorption of the second, as illustrated in Fig. 3.6(a). The real intermediate state A* has a well-defined lifetime, typically 10^{-4} to 10^{-9} s. Because it is excited by a resonantly-absorbed photon, the overall sequential process is referred to as *resonant two-photon excitation*. The second excitation mechanism is *non-resonant two-photon excitation*, in which no resonant intermediate states are involved (Fig. 3.6(b)). A *virtual* state, A*^{v}, is created by the interaction of the first photon with A. Only if a second photon arrives during the duration of the first interaction (about 1 cycle of the electromagnetic radiation, or 10^{-15} s) can it be absorbed. The process is therefore often referred to as *simultaneous* biphotonic absorption, to distinguish it from the sequential mechanism. It is now apparent why high intensities are essential to achieve two-photon excitation. In the resonant process, it is necessary that the second excitation step occurs before loss processes (collisional, intramolecular radiationless, and radiative) depopulate

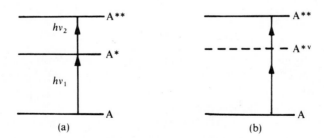

Fig. 3.6 (a) Resonant and (b) non-resonant two-photon absorption.

the intermediate A*, while with the non-resonant process the two photons must reach the absorber within about 10^{-15} s of each other. The resonant process has a higher probability than the non-resonant one, although the two formally distinct mechanisms merge if the photon energy of the exciting radiation in the non-resonant process approaches that of a real intermediate state.

In both the sequential and the simultaneous mechanisms, the essential feature is that the energy of individual photons—still quantized according to the Planck relation—is stored in such a way that the energy of several photons can be used cooperatively. The optical absorbance of the system will be dependent on the incident light intensity, i.e. the Beer–Lambert law (Section 2.4) no longer holds. This behaviour is most obvious in the non-resonant, simultaneous, process involving the virtual intermediate level. The system may be completely transparent at low light intensities, yet absorb radiation at the same wavelength when exposed to high intensities. A good example of absorption by a 'transparent' gas is discussed in Section 5.5 (p. 117): fluorescence in caesium vapour can be excited by intense radiation not corresponding to any (single-photon) spectral transition.

Multiphoton excitation in the ultraviolet region offers the photochemist several interesting opportunities. Excited states not normally accessible can be populated. For example, two-electron excitation can be achieved with two-photon absorption, as can excitation of states with the same parity as the ground state (the selection rules are g ↔ g and u ↔ u, contrary to the rules for single-photon absorption (cf. Section 2.6, p. 28). Of particular importance is the possibility of conducting high-energy photochemistry that is normally associated with ultraviolet radiation of wavelength shorter than, say, 190 nm. This region is referred to as the 'vacuum' ultraviolet, because air absorbs the radiation, so that conventional experiments become increasingly difficult as vacuum techniques have to be adopted. The problem is compounded by the increasing absorbance of window materials (quartz, for example, is limited to $\lambda \simeq 165$ nm: see footnote on p. 155). These problems can be avoided by using multiphoton excitation with wavelengths in the 'conventional' ultraviolet region. A typical example of multiphoton photochemistry in the ultraviolet region is the two-photon dissociation of CH_3I at $\lambda = 193$ nm

$$CH_3I + 2h\nu \rightarrow CH(A^2\Delta) + H_2 + I \qquad (3.37)$$

in which electronically excited CH radicals are formed. Ordinary ultraviolet dissociation of CH_3I yields $CH_3 + I$ as the fragments, and the experiments would be extremely difficult at the single-photon wavelength ($\simeq 86$ nm) corresponding to the biphotonic excitation.

Multiphoton ionization (MPI) is readily achieved with ultraviolet laser radiation. If resonant intermediate states are involved, the process is called *resonance-enhanced* multiphoton ionization (REMPI). Single-photon photoionization for most species requires wavelengths shorter than those transmitted by available window materials, as discussed at the end of Section 3.2. Bi- or multiphoton excitation gives access to ionization studies on a vastly increased range of species. The ions formed can be detected at very low concentrations, and MPI is of great value in multiphoton spectroscopic studies. In addition, there is considerable importance of MPI in mass spectrometry. Photoionization–mass spectrometry with wavelength-selected radiation has long been valued for the specificity with which particular species or particular quantum states can be ionized. The much higher ionization efficiencies, and relative ease of operation, of the laser-MPI technique now greatly extend the usefulness of the technique.

The multiphoton phenomena produced with ultraviolet (or visible) laser radiation are often similar to those observed on single-photon excitation with the appropriate shorter wavelength radiation. Infrared multiphoton excitation, on the other hand, leads to phenomena that would be virtually impossible to study without the use of lasers. Shortly after the development of the carbon dioxide laser (Section 5.7), experiments were performed to see if chemical change could be induced by high intensities of infrared photons. It turns out that the *vibrational photochemistry*, at least of polyatomic molecules, is a very widespread phenomenon. Even though, in most cases, something like 10–40 infrared photons are needed to reach the energy threshold, photofragmentation occurs readily on irradiation of a molecule with a strong vibrational absorption band with intense (pulsed) laser light. For example, the molecule SF_6 may be dissociated by radiation of $\lambda = 10.6 \ \mu m$ from a CO_2 laser

$$SF_6 + nh\nu \rightarrow SF_5 + F \qquad (3.38)$$

in the ν_3 band, even though the SF_5–F bond energy is $348 \ kJ \ mol^{-1}$, or more than 30 times the laser photon energy.

One model of infrared multiphoton dissociation (IRMPD) divides the mechanism into three parts, according to the energy region in the molecule populated by successive absorption events. Figure 3.7 illustrates the mechanism envisaged. In the early stages of excitation, at low energies, vibrational states are discrete (region I). The first photon is absorbed to raise the $v = 0$ to $v = 1$ level in the molecule. Subsequent excitation steps to $v = 2, 3, 4$, etc., must necessarily be out of resonance with monochromatic laser radiation, because the molecular vibrational levels are anharmonic. However, compensating mechanisms (e.g. involving the effect of rotations on the molecular energies or the power-broadening of high-intensity laser lines) do allow non-resonant absorption to occur within this region of sparsely populated vibrational levels. The density of vibrational states increases with energy rapidly in a polyatomic molecule, and at some molecular energy the levels merge to form a 'quasi-continuum' (region II of Fig. 3.7). Within this region, resonant

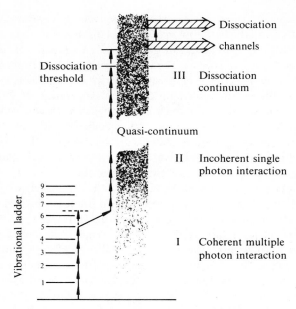

Fig. 3.7 Schematic representation of energy levels in infrared multiphoton dissociation. Coherent multiphoton interaction in the region labelled 'I' explains one way in which absorption can occur in a region of discrete vibrational levels. In the quasi-continuum region, 'II', resonant absorption steps are possible and the excitation can be treated as a stepwise process. The third region, 'III', lies above the dissociation threshold.

absorption processes always exist at the laser wavelength, and a stepwise path is thus available to region III, which lies above the dissociation threshold. Rapid randomization of energy is then followed by unimolecular dissociation to the fragment species.

Multiphoton infrared absorption provides a unique source of molecules with high degrees of internal excitation. A new technique for the study of the problems of unimolecular dissociation is thus available. In addition, IRMPD generally yields products in their electronic ground states, which may well not be the case in conventional single-photon dissociation with ultraviolet or visible radiation. Infrared multiphoton dissociation gives products similar to those obtained by thermal dissociation or pyrolysis, but without the need for high temperatures of the bulk materials. The technique has commanded much attention for the possibilities that it provides in isotope-selective chemistry. Many experiments have repeatedly demonstrated the crucial influence of the first few discrete absorption steps in the overall excitation scheme. Since isotope shifts in vibrational spectroscopy can be relatively large, it is possible to dissociate species containing particular isotopes selectively by tuning the laser to the appropriate $v = 1 \leftarrow v = 0$ transition. Two-laser experiments have shown clearly that a weak, narrow band, laser can be used to pump the first steps in region I, and a high-power laser (whose frequency is relatively unimportant) used to drive the molecule through region II and then to dissociation. For example, UF_6 can be dissociated by pumping the v_3 absorption ($615\,cm^{-1}$) with a weak laser used in conjunction with a more powerful CO_2 laser that on its own does not cause dissociation. The potential applications in isotope separation are apparent, and supplement the techniques that are presented in Section 8.10.

Bibliography

Calvert, J. G. and Pitts, J. N., Jr (1966). *Photochemistry*, Chapters 3–5. John Wiley, Chichester and New York.

Okabe, H. (1978). *Photochemistry of small molecules*, Chapters 2 and 4–7. John Wiley, Chichester and New York.

Simons, J. P. (1971). *Photochemistry and spectroscopy*, Chapters 2 and 3. John Wiley, Chichester and New York.

Lawley, K. P. (ed.) (1985). Photodissociation and photoionization. *Adv. Chem. Phys.* **60**, whole issue.

Gaydon, A. G. (1953). *Dissociation energies and spectra of diatomic molecules*: Chapter IV, 'Photodissociation': Chapter V, 'The Birge–Sponer extrapolation'; Chapter VI, 'Predissociation'. Chapman and Hall, London.

See also Bibliography for Section 7.6

Special topics
Section 3.5

Barltrop, J.A. and Coyle, J. D. (1978). *Principles of photochemistry*, Chapter 3. John Wiley, Chichester and New York.

Turro, N. J. (1978). *Modern molecular photochemistry*, Chapter 6. Benjamin/Cummings, Menlo Park, CA.

Lin, S. H. (ed.) (1980). *Radiationless transitions*, Academic Press, New York.

Freed, K. F. (ed.) (1978). Radiationless transitions in molecules. *Acc. Chem. Res.* **11**, 74.

Formosinho, S. J. (1986). Radiationless transitions and photochemical reactivity. *Pure Appl. Chem.*, **58**, 1173.

Section 3.7: condensed phases

Rice S. A. (1985) Diffusion-limited reactions. In *Comprehensive chemical kinetics* (eds C. H. Bamford, C. F. H. Tipper, and R. G. Compton) Vol. 25. Elsevier, Amsterdam.

Alwattar, A. H., Lumb, M. D., and Birks, J. B., (1973). Diffusion-controlled rate processes. In *Organic molecular photophysics* (ed. J. B. Birks) Vol. 1. John Wiley, Chichester and New York.

Ullman, E. F. (1968). Excited state intermediates in solution photochemistry. *Acc. Chem. Res.* **1**, 353.

Thomas, J. K. (1980). Radiation-induced reactions in organized assemblies. *Chem. Rev.* **80**, 283.

Turro, N. J., Cox, G. S., and Paczkowski, M. A. (1985). Photochemistry in micelles. *Top. Curr. Chem.* **129**, 57.

Ramamurthy, V. and Venkatesan, K. (1987). Photochemical reactions of organic crystals. *Chem. Rev.* **87**, 433.

Section 3.7: dynamic magnetic polarization

Wan, J. K. S. (1980). Theory and applications of chemically induced magnetic polarization in chemistry. *Adv. Photochem.* **12**, 283.

Wan, J. K. S. and Elliot, A. J. (1977). Chemically-induced dynamic magnetic polarization in photochemistry. *Acc. Chem. Res.* **10**, 161.

Symons, M. C. R. and McLauchlan, K. A. (1984). From trapped radicals to transients. *Faraday Discuss. Chem. Soc.* **78**, 7.

Wasielewski, M. R., Norris, J. R., and Bowman, M. K. (1984). Time-domain magnetic resonance studies of short-lived radical pairs in liquid solution. *Faraday Discuss. Chem. Soc.* **78**, 279.

Closs, G. L., Miller, R. J., and Redwine, O. R. (1985). Time-resolved CIDNP: applications to radical and biradical chemistry. *Acc. Chem. Res.* **18**, 196.

Section 3.8, together with other inorganic and organometallic photochemistry

Adamson, A. W. and Fleischauer, P. D. (eds) (1975). *Concepts of inorganic photochemistry*. John Wiley, Chichester and New York.

Lewis, F. D. (1986). Proton-transfer reactions of photogenerated radical ion pairs. *Acc. Chem. Res.* **19**, 401.

Endicott, J. F. (1976). Photochemical reaction pathways of coordination complexes. *Surv. Prog. Chem.* **7**, 41.

Balzani, V., Sabbatini, N., and Scandola, F. (1986). 'Second sphere' photochemistry and photophysics of coordination compounds. *Chem. Rev.* **86**, 319.

Geoffroy, G. L. and Wrighton, M. S. (1979). *Organometallic photochemistry*. Academic Press, New York.

Section 3.9: multiphoton dissociation and ionization

Andrews, D. L. (1986). *Lasers in chemistry*. Springer-Verlag, Berlin.

Steinfeld, J. I. (ed.) (1981). *Laser-induced chemical processes*. Plenum Press, New York.

Ronn, A. V. (1979). Laser chemistry. *Sci. Am.* **240**(5), 102.

Zewail, A. H., Letokhov, V. S. *et al.* (1980). Laser chemistry. *Phys. Today* **33**, November (special issue): includes Laser selective chemistry—is it possible? (A. H. Zewail), Laser-induced chemical processes (V. S. Letokhov).

Donovan, R. J. (1981). Ultraviolet multiphoton excitation: formation and kinetic studies of electronically excited atoms and free radicals, Specialist Periodical Reports. *Gas Kinetics* **4**, 117.

Ashfold, M. N. R. and Hancock, G. (1981). Infrared multiple photon excitation and dissociation: reaction kinetics and radical formation, Specialist Periodical Reports. *Gas Kinetics* **4**, 73.

King, D. S. (1982). Infrared multiphoton excitation and dissociation. *Adv. Chem. Phys.* **50**, 105.

Jortner, J., Levine, R. D., and Rice, S. A. (eds) (1981). Photoselective chemistry. Section 1, Multiphoton-induced photochemistry. *Adv. Chem. Phys.* **48**.

Lupo, D. W. and Quack, M. (1987). I. R. Laser photochemistry. *Chem. Rev.* **87**, 181.

Reisler, H. and Wittig, C. (1985). Multiphoton ionization of gaseous molecules. *Adv. Chem. Phys.* **60**, 1.

Section 3.9: isotope separation

Andrews, D. L. (1986). Lasers in chemistry. Section 5.3. Springer-Verlag, Berlin.

Zare, R. N. (1977). Laser separation of isotopes. *Sci. Am.* **236**(2), 86.

McAlpine, R. D. and Evans, D. K. (1985). Laser isotope separation by the selective multiphoton decomposition process. *Adv. Chem. Phys.* **60**, 31.

Jortner, J., Levine, R. D., and Rice, S. A. (eds.) (1981). Photoselective chemistry. Section 3, One-photon and two-photon photoselective chemistry. *Adv. Chem. Phys.* **48**.

4

Emission processes (1)

4.1 Luminescence

The emission of radiation from excited species is one of the several paths by which the excess energy may be lost (path vi, Fig. 1.1); the general phenomenon of light emission from electronically excited species is known as *luminescence*. In this chapter and the next we shall discuss luminescent processes. First, simple luminescent phenomena are considered, and then, in Chapter 5, *sensitized luminescence* is described: the latter process involves *inter*molecular energy transfer (path iv, Fig. 1.1) and electronic excitation is produced in a species other than the one that was initially excited. *Intra*molecular energy transfer, which populates a different *state* of the same *species*, is discussed in the present chapter.

Luminescent emission provides some of the most reliable information about the nature of primary photochemical processes. Competition exists between emission and other fates of excited species (quenching, reaction, decomposition, etc.), and the dependence of emission intensity on temperature, reactant concentrations, etc., may yield valuable data about the nature and efficiencies of the various processes. In particular, quenching by bimolecular collisions, and unimolecular energy degradation by radiationless transitions, are almost always best studied in terms of their effect on the intensity of luminescence. As well as possessing this fundamental interest, luminescent phenomena are also of considerable importance in several commercial and scientific applications, and an example will be given in Section 8.11.

The various individual luminescent phenomena are named according to the mode of excitation of the energy-rich species. We are concerned primarily with excitation by absorption of radiation, and emission from species excited in this way is referred to as *fluorescence* or *phosphorescence*: the distinction between the two processes is discussed below. Emission following excitation by chemical reaction (of neutral or charged species) is known as *chemiluminescence*, and is described briefly in Section 4.7. Other means of providing electronic excitation, which will not be discussed further, are by heat (e.g. in NO_2—*pyroluminescence*), by an electric field (e.g. in solid ZnS—*electroluminescence*), by electron impact in gases (e.g. in discharge lamps), by electron impact on solid phosphors (e.g. in television tubes—*cathodo-luminescence*), by crushing crystals (e.g. uranyl nitrate—*triboluminescence*), and by rapid crystallization from solution (e.g. strontium bromate—*crystallo-luminescence*). Although we shall have occasion to refer to the luminescence of substances trapped in rigid glasses, we shall omit general discussion of the luminescence of solids. The emission of radiation from solids, especially inorganic compounds, is a complex phenomenon, but of the greatest importance (e.g. colour television requires inorganic phosphors possessing emissions of specific colours and intensities). For an introduction to such luminescence, the reader is referred to Chapter 5 of *Luminescence in chemistry* (Bibliography).

The two emission processes in which the ultimate source of excitation is absorption of radiation—fluorescence and phosphorescence—were orig-inally distinguished in terms of whether or not there was an observable 'afterglow'. That is, if emission of radiation continued after the exciting radiation was shut off, the emitting species was said to be phosphorescent, while if emission appeared to cease immediately, then the phenomenon was one of fluorescence. The essential problem is what is meant by 'immediately' in this context, since the observation of an afterglow will obviously depend not only on the actual rate of decay of the emission (see Section 4.2 for further discussion of emission lifetimes), but also on the techniques used to observe it. Various instruments were devised to observe 'short-lived' luminescence, and in the early 1930s a luminescence with a lifetime of less than about 10^{-4} s was thought to be short-lived and, hence, fluorescent. In 1935 Jablonski inter-preted phosphorescence as being emission from some long-lived metastable electronic state lying lower in energy than the state populated by absorption of radiation (cf. Section 3.5). Several workers (among them Lewis and Kasha, and Terenin) suggested that the long-lived metastable state was, in fact, a triplet state of the species, and as we shall see in Section 4.4, there is now considerable experimental evidence to substantiate this hypothesis. The long lifetime of the emission is a direct consequence of the 'forbidden' nature of a transition from an excited triplet to the ground-state singlet; that electric dipole transitions occur at all where $\Delta S \neq 0$ is due to the inadequacy of S to describe a system in which there is spin–orbit coupling (cf. Section 2.6). Extension of this idea to other systems, not necessarily triplet–singlet, in which $\Delta S \neq 0$ leads to the useful definition of phosphorescence as a *radiative transition between states of different multiplicities*: fluorescence is then under-stood to be a radiative transition between states of the *same* multiplicity. Figure 4.1 is a Jablonski diagram (see pp. 45–47 and Fig. 3.5) showing the processes of fluorescence and phosphorescence. These definitions are used almost universally by organic photochemists, although they might be ex-tended to include within the scope of phosphorescence emission processes involving a transition forbidden by *any* selection rule rather than just the

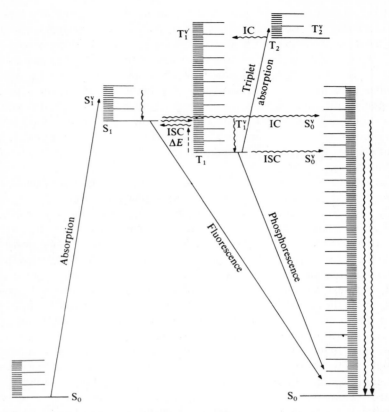

Fig. 4.1 Jablonski diagram showing absorption, and the emission processes of fluorescence and phosphorescence.

$\Delta S = 0$ rule. Since the distinctions between allowed and forbidden transitions are not sharp, the definitions lack some precision.

Absorption of radiation in a singlet–triplet transition is weak, since it is forbidden in the same way as the triplet–singlet phosphorescent emission. It follows that phosphorescence can only be excited inefficiently by direct absorption of radiation, and phosphorescence is much more usually the result of emission from a triplet populated by intersystem crossing from an excited singlet formed on absorption. The sequence of events is illustrated in Fig. 4.1. Absorption populates S_1^v; vibrational energy, at least in condensed phases, is rapidly degraded and S_1^0 can then lose its energy by radiation, intersystem crossing (ISC) to T_1^v, or internal conversion (IC) to S_0^v. It is, perhaps, surprising that ISC to T_1^v, which is spin-forbidden by a radiationless transition selection rule, can compete effectively with spin-allowed fluorescence and IC to S_0^v; phosphorescence is, however, observed in many systems, suggesting that IC from $S_1 \rightsquigarrow S_0$ is relatively inefficient. A complete understanding of the photochemistry of a molecule really requires that the efficiencies (i.e. quantum yields) be known for all the processes occurring. Even if chemical reaction, decomposition, and physical quenching of an excited species do not occur, it is still necessary to measure quantum yields for fluorescence (ϕ_f), phosphorescence (ϕ_p), intersystem crossing $T_1 \rightsquigarrow S_0$ (ϕ_{ISC}) and internal conversion $S_1 \rightsquigarrow S_0$ (ϕ_{IC}). With the restrictions on the processes occurring, it follows that

$$\phi_f + \phi_p + \phi_{ISC} + \phi_{IC} = 1 \tag{4.1}$$

(although the relative magnitudes of the four quantum yields may be affected by the external environment).

4.2 Kinetics and quantum efficiencies of emission processes

Considerable information about the efficiencies of radiative and radiationless processes can be obtained from a study of the kinetic dependence of emission intensity (or quantum yield) on concentrations of emitting and quenching species. In this section we shall consider first the application of stationary-state methods to fluorescence (or phosphorescence) quenching, and then discuss the lifetimes of luminescent emission under non-stationary conditions.

Observable effects in the quenching of fluorescence are usually the result of competition between radiation and bimolecular collisional deactivation of *electronic* energy, since vibrational relaxation is normally so rapid, especially in condensed phases, that emission derives almost entirely from the ground vibrational level of the upper electronic state: this point is discussed further in the next section. The simplest excitation–deactivation scheme, which does not allow for intramolecular radiationless processes, is

rate:

$$X + h\nu_{abs} \rightarrow X^* \qquad \text{absorption} \qquad I_{abs} \tag{4.2}$$

$$X^* + M \xrightarrow{k_q} X^{\cdot} + M \quad \text{quenching} \qquad k_q[X^*][M] \tag{4.3}$$

$$X^* \rightarrow X + h\nu_{em} \qquad \text{emission} \qquad A[X^*] \tag{4.4}$$

Solution of the steady-state equations for $[X^*]$ (i.e. with $d[X^*]/dt = 0$) leads to the result that

$$I_{emitted} = A[X^*] = \frac{AI_{abs}}{A + k_q[M]} \tag{4.5}$$

(note that the rate constant A is the Einstein coefficient for spontaneous emission). Equation (4.5) can be inverted to give the *Stern–Volmer* relation

$$\frac{1}{I_{\text{emitted}}} = \frac{1}{I_{\text{abs}}}\left(1 + \frac{k_q}{A}[M]\right) \tag{4.6}$$

Figure 4.2 shows a typical plot of $1/I_{\text{emitted}}$ against $[M]$ for the quenching of the fluorescence of an aqueous quinine sulphate solution by the chloride ion. The values of the slope and intercept can be used to give a value of k_q/A even if I_{emitted} is measured in arbitrary units and I_{abs} is not determined. Thus, if the Einstein A factor is known, or can be measured, the value of the quenching rate constant can be calculated. The A factor can be calculated from the B factor by use of the ν^3 relationship presented as eqn (2.1) (p. 19) (and B itself can, of course, be calculated from the measured integrated extinction coefficient for the absorption band, as described in Section 2.4). It is also possible,

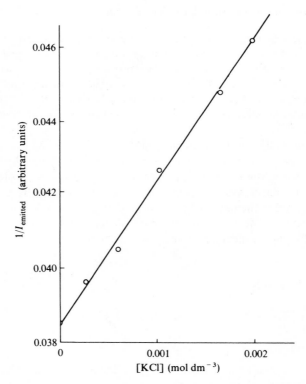

Fig. 4.2 Stern–Volmer plot for the quenching of quinine sulphate fluorescence by the chloride ion. (Averaged points from practical course experiment, Physical Chemistry Laboratory, Oxford.)

under suitable conditions, to measure A directly by o
of emission after suddenly extinguishing the illumina
see later in this section, the fluorescence or phosphore
shorter than the 'natural' radiative lifetime as a result
intramolecular non-radiative energy degradation, so
taken in the interpretation of emission decay measurem
mentioned in Section 4.6, the observed lifetime may
natural radiative lifetime.)

Rate constants for quenching can be compared with t ~.............. ~j ~..~
collision theory of chemical kinetics. According to this theory, a rate constant, k, is given by

$$k = \sigma_{\text{coll}}^2 \left(\frac{8\,\mathbf{k}T}{\pi\mu}\right)^{1/2} e^{-E_a/RT} \tag{4.7}$$

(σ_{coll} is the *collision cross-section*, equal to $\pi(r_A + r_B)^2$, where r_A, r_B are the gas-kinetic collision radii of the reaction partners, and μ is their reduced mass; E_a is the activation energy for the reaction). E_a is expected to be near zero for collisional quenching, so one way of making the comparison is to calculate, from k_q, the *quenching cross-section* (which we will write as σ_q^2) and compare it with σ_{coll}^2. Table 4.1 shows some data obtained for the quenching of the fluorescence of NO together with values of the gas-kinetic cross-section. The ratio $\sigma_q^2/\sigma_{\text{coll}}^2$ corresponds to the familiar P factor of the collision theory (assuming that $E_a = 0$); the results suggest that quenching efficiency increases with increasing complexity of M (note especially the effectiveness of CO_2, which appears to quench somewhat faster than the collision rate). Even for

Table 4.1 *Cross-sections for the quenching of NO $(A^2\Sigma \rightarrow X^2\Pi)$ fluorescence (Data quoted by Okabe, H. (1978).* Photochemistry of small molecules, *Table V-6. Wiley, New York).*

Quenching gas M	$10^{16}\sigma_q^2$	$10^{16}\sigma_{\text{coll}}^2$	$\dfrac{\sigma_q^2}{\sigma_{\text{coll}}^2}$
	(cm^2 molecule^{-1})		
He	1.3	27.9	0.05
Ar	<0.014	36.7	$\simeq 0$
N_2	0.014	41.2	$\simeq 0$
CO	3.3	40.5	0.08
O_2	29	38.7	0.75
NO	37	39.8	0.93
CO_2	68	43.7	1.56

...le, only about 20 gas-kinetic collisions are needed, on average, to bring ...out quenching.

The relatively great rates of the quenching process may mean that in solution the rate is determined more by the rate of diffusion of the quenching and emitting molecules than by the rate of collision. An *approximate* expression for the diffusion-limited rate constant, k_{diff}, is given by the Debye equation

$$k_{diff} \sim \frac{8RT}{3\eta} \times 10^3 \, dm^3 \, mol^{-1} \, s^{-1} \tag{4.8}$$

where η is the viscosity of the solvent in newton-seconds per square metre and $R = 8.3 \, JK^{-1} \, mol^{-1}$. For water at room temperature, $\eta \sim 10^{-3} \, Nsm^{-2}$, so that $k_{diff} \sim 10^{10} \, dm^3 \, mol^{-1} \, s^{-1}$. The results for the quenching of quinine sulphate fluorescence (Fig. 4.2) give a value for $k_q/A \sim 100 \, dm^3 \, mol^{-1}$. For the transition involved, $A \sim 4.3 \times 10^7 \, s^{-1}$, so that k_q approaches the diffusion-controlled limit. Note that, since η is temperature-dependent, k_q may increase with temperature so that there *appears* to be an activation energy for the process; however, the true E_a, in the sense of the energy needed for reaction once a collision occurs, can still be zero. Although the quenching rate approaches the diffusion-controlled limit, it is not necessarily true that every molecular *collision* leads to deactivation. The diffusive process limits the rate at which the excited species and the quencher come together, but prolongs each *encounter* so that several hundred collisions are possible before the two species diffuse apart (see also Section 3.7).

The kinetics of the emission process has been developed in terms of excitation, emission, and collisional deactivation steps. If intramolecular energy-loss processes take place, then additional first-order terms must be added to the denominator of eqn (4.5). Thus, if the rate constant that describes first-order energy degradation is k_1, the modified form of eqn (4.5) is

$$I_{emitted} = \frac{AI_{abs}}{A + k_1 + k_q[M]} \tag{4.9}$$

In this case a Stern–Volmer plot ($1/I_{emitted}$ vs. [M]) will have an intercept of $(1 + k_1/A)I_{abs}^{-1}$ and slope $(k_q[M]/A)I_{abs}^{-1}$, and k_q/A cannot be determined unless $I_{emitted}$ and I_{abs} are measured. Only the *ratio* $I_{emitted}/I_{abs}$ is needed, so that *absolute* values of intensity are not necessary, although arbitrary scales must be corrected to relative photon (rather than energy) fluxes if the wavelength distributions of exciting and emitted radiation are not identical. The ratio $I_{emitted}/I_{abs}$ is, of course, the quantum yield, ϕ_1, for the luminescent process (ϕ_f, fluorescence quantum yield; ϕ_p, phosphorescence quantum yield), and the intercept of a modified Stern–Volmer plot of $1/\phi_1$ vs. [M] will indicate the relative ratio of radiative and radiationless loss processes. Hence, ideally, it is possible to determine the rate of internal conversion or inter-system crossing in luminescent systems.

If a species exhibits both fluorescence and phosphorescence, and the phosphorescent state T_1 is excited by intersystem crossing from S_1, we derive the quantum yields from the following reaction scheme:

$$S_0 + hv \longrightarrow S_1 \qquad \text{excitation} \qquad (4.10)$$

$$S_1 + M \xrightarrow{k_1} S_0 + M \qquad \text{fluorescence quenching} \qquad (4.11)$$

$$S_1 \xrightarrow{A'} S_0 + hv' \qquad \text{fluorescence} \qquad (4.12)$$

$$S_1 \xrightarrow{k_2} T_1 \qquad \text{ISC} \qquad (4.13)$$

$$S_1 \xrightarrow{k_3} S_0 \qquad \text{IC} \qquad (4.14)$$

$$T_1 + M \xrightarrow{k_4} S_0 + M \qquad \text{phosphorescence quenching} \qquad (4.15)$$

$$T_1 \xrightarrow{A''} S_0 + hv'' \qquad \text{phosphorescence} \qquad (4.16)$$

$$T_1 \xrightarrow{k_5} S_0 \qquad \text{ISC} \qquad (4.17)$$

Solution of the stationary-state equations for $[S_1]$ and $[T_1]$ leads to the results

$$\phi_f = \frac{A'}{A' + k_2 + k_3 + k_1[M]} \qquad (4.18)$$

and

$$\phi_p = \frac{A''}{A'} \cdot \left(\frac{k_2}{A'' + k_5 + k_4[M]} \right) \qquad (4.19)$$

Measurement of both ϕ_f and ϕ_p under conditions where bimolecular quen-ching is negligible (or, alternatively, extrapolation to $[M] = 0$) can thus lead to values for $(k_2 + k_3)$ and k_5/k_2 if A' and A'' are known. Direct measurement of the rates of process (4.13) have been made in some cases, and the values of k_2 obtained suggest that $k_3 \ll k_2$ (see pp. 79–80); with k_2 known, k_5 may be estimated.

The discussion of the preceding paragraph indicates how kinetic data and measurements of quantum yields might ideally be used to assess the impor-tance of the several photochemical processes (4.10–4.17). It is, however, in the nature of steady-state kinetic calculations that *ratios* of rate constants are obtained: for example, the expressions for the intensities or quantum yields in eqns (4.6), (4.9), (4.18) and (4.19) all involve ratios of rate constants to the Einstein A factor for emission. Individual rate constants can often be determined from a comparison of kinetic data obtained under stationary conditions with those obtained under non-stationary conditions. For the present purposes, the non-stationary experiment often involves determination of fluorescence or phosphorescence life-times (τ_f, τ_p). If a process follows first-order kinetics described by a rate constant k, the mean lifetime, τ (the time taken for the reactant concentration to fall to $1/e$ of its initial value), is given by

$$\tau = 1/k \qquad (4.20)$$

If the loss of luminescent species after the exciting radiation is shut off is unimolecular or pseudo-first-order, then we may define mean lifetimes for the decay of emission as the inverse of the sum of all the effective first-order rate constants. Thus, for the very simple reaction scheme (4.2)–(4.4)

$$\tau = (A + k_q[M])^{-1} \qquad (4.21)$$

Measurement of τ as a function of [M] and extrapolation to [M]=0 then yields a value for A; since A can also be calculated, via B, from extinction coefficients in absorption, the measured value may afford a check on the calculated one (cf. Section 4.6). In general, when [M]$\neq$0, the observed lifetime is shorter than the 'natural' radiative lifetime ($=1/A$). Where intramolecular loss processes occur, and the stationary-state kinetics are described by eqn (4.9), then

$$\tau = (A + k_1 + k_q[M])^{-1} \qquad (4.22)$$

If a reliable value of A may be calculated from B, then k_1 may be determined explicitly from τ without the need for absolute quantum yield determinations. Most quoted rate constants for IC or ISC processes derive, in fact, from lifetime measurements.

4.3 Fluorescence

An electronically excited atom must lose its energy either by emission of radiation or by collisional deactivation: chemical decomposition is not possible, and radiationless degradation (involving an increase in translational energy) is extremely improbable. At low enough pressures, therefore, fluorescent emission is expected from all atoms. Many molecular species, however, either do not exhibit fluorescence or fluoresce weakly even when bimolecular reaction or physical deactivation does not occur. Some general principles can suggest whether a polyatomic organic molecule is likely to be strongly fluorescent. First, absorption must occur at a wavelength long enough to ensure that chemical dissociation does not take place. Absorption to an unstable state is clearly very unlikely to result in fluorescence. Further, in many molecules in which the absorption maximum corresponds to an energy greater than the cleavage energy of the least stable bond, no fluorescence is observed. Secondly, intramolecular energy transfer must be relatively slow compared to the rate of radiation. This appears to mean that ISC from $S_1 \leadsto T_1$ must be slow (we have mentioned in Section 4.2 and shall discuss later in this section the inefficiency of the IC $S_1 \leadsto S_0$ process); in Section 4.5 we shall see that ISC is normally slower for $^1(\pi, \pi^*) \leadsto {}^3(\pi, \pi^*)$ states than for ISC involving n, π^* states, and that the efficiency of the process increases as the energy separation of S_1 and T_1 decreases. Experimental observations of fluorescence are in accord with these ideas: the simpler carbonyl compounds, in which the longest absorption corresponds to an $n \rightarrow \pi^*$ transition, are rarely fluorescent (but often phosphorescent) while aromatic hydrocarbons ($\pi \rightarrow \pi^*$ absorption) are frequently fluorescent. Increasing conjugation in hydrocarbons shifts the first ($\pi \rightarrow \pi^*$) absorption maximum towards longer wavelengths, and thus increases the probability of fluorescence rather than decomposition. High ring density of the π electrons also seems important for high fluorescence yields. Geometrical factors such as rigidity and planarity also affect the efficiency of fluorescence. For example, *cis*-stilbene, C_6H_5–CH=CH–C_6H_5, is non-fluorescent in ordinary solutions, while the more rigid derivative 1,2-diphenylcyclobutene, C_6H_5–☐–C_6H_5, has a quantum yield for fluorescence approaching unity. Even *cis*-stilbene itself becomes appreciably fluorescent in highly viscous solvents. The radiationless $S_1 \leadsto S_0$ internal conversion, which would deactivate the fluorescing S_1 state, is inhibited with rigid or constrained molecules; an explanation of the effect will be presented in Section 4.5.

The simplest type of fluorescence is *resonance fluorescence*, in which the radiation emitted is of the same wavelength as the exciting radiation. Resonance fluorescence is observed only in the gas phase at low pressures, and only with atoms or simple molecules. For example, in I_2 vapour, at a pressure around 10^{-2} mmHg, resonance fluorescence is observed on excitation by light of suitable wavelength. Absorption of monochromatic light, of wavelength corresponding to a specific vibrational transition ($v', 0$) in the $^3\Pi \leftarrow {}^1\Sigma_g^+$ system of I_2, populates exclusively the v' level of the upper state and radiation from that state gives rise to the resonance fluorescence (see Fig. 3.1). Transitions also occur from v' to v'' levels higher than zero, so that a progression of bands is observed at wavelengths longer than the exciting wavelength,[†] in accordance with an empirical observation of Stokes: the lines

---

[†]The bands are often referred to as part of the resonance emission spectrum since they derive from the level of v' initially populated. Properly speaking, however, the resonance line is the same wavelength as the exciting line.

are called *Stokes lines*. (Much weaker *anti-Stokes lines* are observed at shorter wavelengths than the exciting wavelength, and result from absorption from $v'' > 0$ and fluorescence to lower v'' levels: thermal populations of $v'' > 0$ are small and the anti-Stokes lines are correspondingly weak.) Irradiation by polychromatic light can obviously excite many v' levels, and fluorescent emission can then observed from all these levels, up to the dissociation limit (in I_2, up to $\lambda = 499$ nm).

Radiationless transitions can lead to a rapid depopulation of v' levels near the crossing point of potential energy curves: this depopulation is one of the reasons why resonance fluorescence of complex molecules is rare, even at low pressure. As was pointed out in Section 3.3, if the crossing process occurs only ten times more rapidly than radiation (e.g. if, for an ordinary 'allowed' fluorescence, the rate constant for crossing is about 10^9 s^{-1}), then the emitted intensity will be reduced by a factor of about ten, and 'breaking-off' in emission is a sensitive test of predissociation. Nitrogen dioxide fluorescence illustrates this phenomenon nicely. The primary quantum yield for decomposition of NO_2 increases sharply at wavelengths shorter than those at which the absorption spectrum becomes diffuse ($\lambda < 400$ nm approximately). The quantum yield for fluorescence in NO_2 shows the reverse trend, being intense at wavelengths longer than about 410 nm, and negligible for $\lambda \lesssim 390$ nm. The *sum* of fluorescence and dissociation quantum yields is unity at all wavelengths from about 360 nm to 450 nm. Electronically excited NO_2 can also be formed chemically by the reaction

$$O + NO + M \rightarrow NO_2^* + M \tag{4.23}$$

(see Section 4.7), and the short wavelength limit of chemiluminescent emission from reaction (4.23) corresponds exactly to the wavelength of onset of diffuseness in absorption.

Stepwise collisional relaxation of vibrational excitation is a relatively efficient process, cross-sections for single-quantum deactivation being between 1% and 100% of the gas-kinetic cross-section for many quenching gases. Resonance fluorescence is not expected, therefore, at pressures at which the kinetic collision frequency greatly exceeds the spontaneous emission rate, and, for $A \sim 10^8$ s^{-1}, observation of resonance emission is confined to pressures at least below about 1 mmHg (and less, if A is smaller than 10^8 s^{-1}). Lower vibrational levels of the upper electronic state are populated from the v' level produced on absorption, and at moderate pressures, at which emission and vibrational quenching still compete, emission may be observed from all vibrational levels in the upper state up to v'. For example, although individual bands are not seen, the low-pressure fluorescence of NO_2 shifts to progressively longer wavelengths as the total pressure in the system is increased.

At higher gas pressures, at which the collision rate greatly exceeds the rate of emission, vibrational relaxation is essentially complete, and no fluorescence is observed from $v' > 0$. Vibrational relaxation is extremely probable in solution, and fluorescence from vibrationally excited levels is never observed in the liquid phase. Furthermore, neither the fluorescence spectrum nor deactivation rates are affected by changes in exciting wavelength so long as it lies within the absorption bands. $S_0 \rightarrow S_1$ transitions in organic compounds are often partially forbidden; to obtain sufficient light absorption to render gas-phase fluorescence detectable frequently requires large pressures, which result in vibrational relaxation to $v' = 0$. This relaxation, together with the probablility of radiationless loss in complex species, accounts for the rarity of resonance or vibrationally hot emission phenomena in organic molecules.

The intensity of each vibrational emission band depends, in the same way as absorption intensities, on the operation of the Franck–Condon principle. Simple diatomic species frequently have greatly different internuclear separations in ground and excited states: Fig. 2.3(b) is representative of the type of situation for O_2, where the upper state is larger than the lower state. Absorption in O_2 originates almost entirely from $v'' = 0$, and is most probable to v' levels from 7 to 11, and $\lambda_{max} \sim 185$ nm, while fluorescence, at pressures at which vibrational relaxation is complete, is strongest around the (0, 14) band at $\lambda \sim 340$ nm. In contrast, the (0, 0) band is the most intense for many organic molecules, and the maxima of intensity both in absorption and in emission therefore correspond to the same transition. This observation suggests that upper and lower electronic states of such molecules must be of similar size and shape, and it is likely that the vibrational *spacings* will be the same in both states. Figure 4.3 shows emission and absorption spectra of a solution of anthracene in benzene. The two spectra are almost mirror images of each

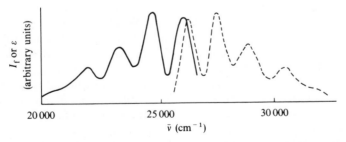

Fig. 4.3 Emission and absorption spectra of a solution of anthracene in benzene, plotted on a frequency scale to reveal the energy spacing of vibrational bands. The emission spectrum is shown as a solid line, and the absorption spectrum as a broken line. (From Bowen, E. J. (ed.) (1968). *Luminescence in chemistry*, p. 10. Van Nostrand, Princeton, NJ.)

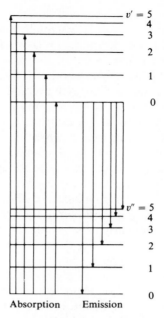

Fig. 4.4   Two electronic states with similar vibrational spacings: the absorption and
emission bands bear a mirror-image relationship to each other.

other on the wavenumber (i.e. energy), rather than wavelength, scale employed. Figure 4.4 illustrates the energy levels in upper and lower states:
because spacings are similar, the (0, 1) emission band will lie at the same
energy below the (0, 0) band as the (1, 0) absorption band lies above it, and so
on. This 'mirror-image' relationship is of frequent occurrence in the fluorescence of organic substances; assumption of its existence is sometimes useful
in sorting out overlapping emission spectra. Whether or not there is a mirror
image, the spacing of emission bands indicates the vibrational levels in the
ground electronic state, while the spacing of absorption bands depends on
vibrational spacing in the upper state.

   The (0, 0) bands in Fig. 4.3 are seen to lie at slightly different wavelengths in
absorption and in emission; Fig. 4.5 shows a more pronounced separation of
(0, 0) bands for dimethylnaphtheurhodine in two solvents. The separations
are caused by energy loss to the solvent environment. The equilibrium
interactions with the solvent may be different for ground and excited states of
the solute (these are mainly electrical interactions, via the dipole moment of
the solute if sizes are similar in both states: cf. Section 6.2). Although the
species cannot relax to the equilibrium interaction energy during the absorption process (i.e. in about 10^{-15} s), it can do so before fluorescent emission

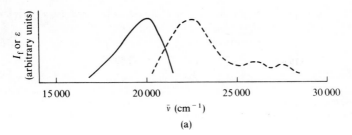

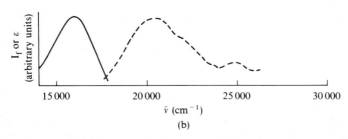

Fig. 4.5 Fluorescence (solid line) and absorption (broken line) bands for solutions of dimethylnaphtheurhodine in (a) hexane, (b) ethanol. There is a pronounced separation of the (0, 0) bands (intensity maxima). (From Bowen, E. J. (ed.) (1968). *Luminescence in chemistry*, p. 10. Van Nostrand, Princeton, NJ.)

occurs (i.e. in about 10^{-8} s). Thus the energy of 'equilibrium $v' = 0$' is rather lower than that of $v' = 0$ populated by absorption, and there is separation of (0, 0) bands. The magnitude of the separation depends on the dipole moment of the excited state of the emitting species, and also on the polarity of the solvent (cf. Fig. 4.5 for hexane and ethanol as solvents). Measurement of (0, 0) band separations is, in fact, used to estimate dipole moments of excited species. At very low temperatures the separations become small because molecular movements are frozen, and the separations may also be small at very high temperatures because of violent motions.

Organic fluorescence usually originates from the *lowest excited singlet* level, S_1, even though absorption may initially populate a higher singlet (e.g. $S_2, S_3, \ldots S_n$). Apparently there is rapid internal conversion from S_n to S_1, followed by vibrational degradation (possibly the process involves intersystem crossings via intermediate triplet states). Internal conversion $S_1 \rightsquigarrow S_0$ must be much slower than $S_n \rightsquigarrow S_1$, since emission from S_1 can compete with radiationless loss, and perhaps more tellingly, since forbidden intersystem crossing $S_1 \rightsquigarrow T_1$ can compete and lead to phosphorescent emission. Measurements of $\phi_{ISC(S_1 \to T_1)}$ and of ϕ_f for many aromatic molecules suggest

Table 4.2   *Quantum yields for fluorescence* $(S_1 \rightarrow S_0 + h\nu)$ *and intersystem crossing* $(S_1 \rightsquigarrow T_1)$ *for some aromatic hydrocarbons in ethanol solution (Data from Birks, J. B. (ed.)* *(1975). Organic molecular photophysics,* Vol. 2, *Tables 2.6 and 3.4. Wiley, London)*

Compound	ϕ_f	ϕ_{ISC}	$\phi_f + \phi_{ISC}$
Benzene	0.04	0.15	0.19
Naphthalene	0.80	0.21	1.01
Fluorene	0.32	0.68	1.00
Anthracene	0.72	0.32	1.02
Tetracene	0.66	0.16	0.82
Phenanthrene	0.85	0.13	0.98
Pyrene	0.38	0.65	1.03
Chrysene	0.85	0.17	1.03

that direct internal conversion $S_1 \rightsquigarrow S_0$ is often unimportant. Table 4.2 shows values of ϕ_f and ϕ_{ISC} for some aromatic hydrocarbons in ethanol solution.

In most cases, the sum of ϕ_f and ϕ_{ISC} is close to unity, so that the quantum yield for internal conversion, $\phi_{IC} = 1 - (\phi_f + \phi_{ISC})$, must be zero within experimental error. Only for benzene itself, and for tetracene (naphthacene), does IC appear to be important for the compounds listed. Since the IC process is *competitive* with radiation or ISC, it is *relatively* slow. We shall see, in Section 4.5, that the efficiency of intramolecular energy transfer usually decreases as the energy difference between two levels increases, and it is possible that the inefficiency of the $S_1 \rightsquigarrow S_0$ process, especially compared with that of $S_n \rightsquigarrow S_1$, reflects the relatively large energy separation between S_1 and S_0.

The observation that fluorescent emission occurs only from S_1, and not from higher singlet states, is common in organic photochemistry, and Kasha has enunciated a rule that *the emitting electronic level of a given multiplicity is the lowest excited level of that multiplicity.* One organic compound whose fluorescence is an exception to this rule is azulene. In this molecule, the $S_2 - S_1$ gap is relatively large, so that the normally rapid $S_2 \rightsquigarrow S_1$ conversion is slowed down, and fluorescence is mainly of the $S_2 \rightarrow S_0 + h\nu$ transition. Many thioketones (R—C—R′) behave in a similar manner.
$$\overset{\displaystyle \|}{\underset{\displaystyle S}{}}$$

Because fluorescent emission in organic compounds comes predominantly from the lowest vibrational level of the lowest excited singlet state of the molecule, it is often found that the fluorescence quantum yield is independent of the wavelength of exciting radiation. Since

$$I_f = I_{abs}\phi_f = I_0\phi_f(1 - e^{-\alpha cd}) \tag{4.24}$$

at very low concentrations, where $\alpha cd \ll 1$,

$$I_f = I_0 \phi_f \alpha cd \qquad (4.25)$$

It follows from the constancy of ϕ_f with exciting wavelength that I_f is proportional to α at any wavelength if the incident intensity is the same at all wavelengths. That is, the *fluorescence excitation spectrum* is the same as the absorption spectrum in sufficiently dilute solutions. This result forms the basis of *spectrofluorimetry*. Figure 4.6 shows part of the absorption spectrum (a) of 1,2-benzanthracene in ethanol and the fluorescence excitation spectrum (b) of a much more dilute solution. The technique clearly makes it possible to obtain 'absorption' spectra at very low solute concentrations (typically, at concentrations as low as 10^{-9} mol dm^{-3}), and the great sensitivity of spectro-fluorimetry makes it a useful analytical tool. It should be noted that the structure of a fluorescence *emission* spectrum, as distinct from the *excitation* spectrum, is often not sufficiently specific in condensed phases for molecular identification to be possible.

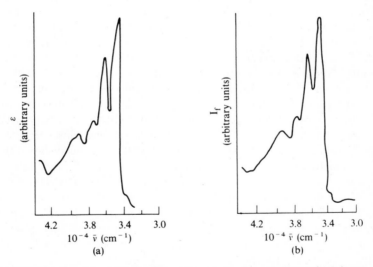

Fig. 4.6 Absorption (a) and fluorescence (b) spectra for solutions of 1,2-benzanthra-cene in ethanol. (From Parker, C. A. (1958). *Nature* **182**, 1002.)

4.4 Phosphorescence

Organic molecules trapped in rigid glassy media have long been known often to show a phosphorescent afterglow following irradiation by light. It is now understood that phosphorescence in organic molecules is emission of a 'forbidden' band, usually from a triplet level. Because of the long radiative lifetime of such transitions, collisional deactivation of the triplet competes effectively with radiation, and visible phosphorescence is not normally observed unless the collisional deactivation rate is sufficiently reduced. In rigid media, species are unable to diffuse towards each other, and bimolecular deactivation is slow. The earliest investigations of phosphorescence employed solutions of dyes in gelatin, and subsequently in boric acid glass at room temperature. More satisfactory rigid media are now used: mixtures of ether, isopentane, and ethanol (EPA) frozen at liquid nitrogen temperature (77 K) are frequently employed, and thin films of various plastics are becoming popular as rigid matrices. The highest purity of the solvents is necessary to avoid swamping the phosphorescence of the solute by luminescence of the impurities.

Although the first observations of phosphorescence were confined to rigid glasses, it was soon appreciated that phosphorescence could appear in other phases. Emission from biacetyl vapour is one of the best-known examples of gas-phase phosphorescence. Fluid solutions of species that are phosphorescent in low-temperature glasses also generally show emission, so long as the radiationless transitions from T_1 to S_0 do not show an increased rate at the higher temperatures. It is, of course, essential that the solvent does not deactivate the triplet, and that quenching impurities are rigorously excluded. Residual impurities may still make the emission intensity weak, and artificially reduce the luminescence lifetime. Perfluoroalkanes make suitable solvents for the study of phosphorescence at room temperature.

Confirmation that the emitting species in phosphorescent organic molecules is a triplet has come from several sources. In the 1940s it was discovered that a solution of fluorescein in boric acid glass became paramagnetic under intense irradiation; more recently it has been shown that the paramagnetism and the phosphorescence decay at identical rates when irradiation ceases. The electron paramagnetic resonance (EPR) technique is capable of detecting triplet species. The first unambiguous detection of a triplet by EPR was of the $\Delta M = \pm 1$ transition in an irradiated single crystal of naphthalene in durene; $\Delta M = \pm 2$ transitions have also been observed in irradiated naphthalene. Triplet concentrations, measured by EPR, in solid solutions of certain phosphorescent aromatic ketones decay, after irradiation, at the same rate as the phosphorescence.

Optical absorption to a higher triplet has afforded further evidence that the emitting state in phosphorescence is a triplet. Intense irradiation of a boric acid glass containing fluorescein leads to the appearance of a new absorption band due to triplet–triplet absorption. Flash photolysis (see Chapter 7), in which a sample is exposed to a brief, intense flash of light, can be used to produce high transient concentrations of triplet species: kinetic absorption spectroscopy of the system enables the build-up and decay of several singlet and triplet levels to be followed as a function of time.

We must now give some consideration to the nature of the forbidden triplet–singlet emission. In Section 2.6 we suggested that electric dipole transitions could occur with $\Delta S \neq 0$ if S did not offer a good description of the system. Optical transitions between triplet and singlet states can take place if the triplet has some singlet character, or vice versa. In organic molecules some 'mixing' of singlet and triplet states takes place as a result of a small amount of spin–orbit interaction. It turns out that the spin–orbit perturbation is forbidden between states of the same configuration, so that, for example, a $^3(\pi, \pi^*)$ state must 'borrow' its singlet character from $^1(n, \pi^*)$ and $^1(\sigma, \pi^*)$ states rather than from $^1(\pi, \pi^*)$. Similarly, a $^3(n, \pi^*)$ state mixes with a $^1(\pi, \pi^*)$ state. Since radiative transitions from $^1(\pi, \pi^*)$ states to the ground state are fully allowed, while transitions from $^1(n, \pi^*)$ are, in general, somewhat forbidden, it follows that $T(n, \pi^*) \rightarrow S_0$ transitions are more allowed than $T(\pi, \pi^*) \rightarrow S_0$. Thus, the relative probability of triplet–singlet transitions from (n, π^*) and (π, π^*) states is opposite to that observed for singlet–singlet transitions. Experimental determinations of natural phosphorescence lifetimes agree with the predictions: in aromatic hydrocarbons having a $^3(\pi, \pi^*)$ state for T_1 the radiative lifetime is roughly 1–10 s, while for carbonyl compounds possessing a lowest triplet state of $^3(n, \pi^*)$ character the lifetime is usually $10^{-2} - 10^{-1}$ s.

Absorption leading to direct population of an excited triplet state from the singlet ground state is weak because the transition is forbidden: decadic molar extinction coefficients may be as low as 10^{-5}. However, in some cases it has proved possible to excite phosphorescence by irradiation with light absorbed in the $T_1 \leftarrow S_0$ system. Just as with fluorescence, there is often a mirror-image relationship between absorption and phosphorescence spectra. It would appear, therefore, that in the relatively large organic molecules the vibrational spacings are nearly identical in all three lowest states (S_0, T_1, and S_1). The (0, 0) separations in the $T_1 \leftarrow S_0$ absorption and emission spectra are, however, relatively large (~ 500 cm^{-1}) as a result of slight conformational differences between ground and excited states. Hence, triplet energies based solely on presumed (0, 0) band maxima in emission only partly represent the energetics of an absorbing system. Good absorption spectra of $T_1 \leftarrow S_0$ transitions are difficult to obtain by ordinary techniques, but the weakness of the absorption makes it possible to use the *phosphorescence excitation spectrum* to determine the absorption spectrum (this is spectrophosphorimetry—cf. Section 4.3 for spectrofluorimetry).

Phosphorescence most commonly follows population of T_1 via intersystem crossing from S_1, itself excited by absorption of light. The T_1 state is usually of lower energy than S_1, and the long-lived (phosphorescent) emission is almost always of longer wavelength than the short-lived (fluorescent) emission. The relative importance of fluorescence and phosphorescence depends on the rates of radiation and intersystem crossing from S_1; the absolute quantum yields depend also on intermolecular and intramolecular energy-loss processes, and phosphorescent emission competes not only with collisional quenching of T_1 but also with intersystem crossing to S_0. The difference between the overall rate of triplet production from S_1 and the rate of phosphorescent emission can be used to calculate the efficiency of the $T_1 \rightsquigarrow S_0$ process under conditions in which bimolecular quenching is negligible.

4.5 Radiationless transitions in complex molecules: intramolecular energy transfer (2)

Now that the fundamental principles of fluorescent and phosphorescent emission have been given, the general topic of intramolecular energy transfer can be treated in greater detail.

Selection rules for radiationless transitions in simple molecules have been given in Section 3.3 in the discussion of predissociation. It is more difficult to develop similar rules for complex species, but at least the spin rule, $\Delta S = 0$, is still of importance. As for emission processes, the occurrence of transitions with $\Delta S \neq 0$ is a result of spin–orbit coupling in the molecule, and the transition probabilities for intersystem crossing follow virtually the same pattern as those discussed in the last section for radiative processes. On the basis of these ideas, El-Sayed has suggested the following 'rules' for spin-forbidden intramolecular energy transfer:

$$^{1 \text{ or } 3}(n, \pi^*) \leftrightarrow ^{3 \text{ or } 1}(\pi, \pi^*); \; ^3(n, \pi^*) \nleftrightarrow ^1(n, \pi^*); \; ^3(\pi, \pi^*) \nleftrightarrow ^1(\pi, \pi^*) \quad (4.26)$$

These considerations are concerned with the electronic component in the probability of particular types of radiationless transition. Experimental evidence, some of which will be presented later, shows that the probability for energy transfer is also some inverse function of the energy gap between the two states for a given type of electronic transition. This result can be understood in terms of the operation of the Franck–Condon principle in radiationless transitions. The principle was discussed for *radiative* transitions in Section 2.7. It proposes that the nuclei of a molecule do not move during the course of an electronic transition: i.e. the transitions are 'vertical' on a potential energy diagram (cf. Fig. 2.3(a) and (b)). In an intramolecular radiationless transition, the sum of the electronic and the vibrational energies must remain constant, in distinction to the radiative case where the photon provides or removes the energy difference between starting and finishing states. In the radiationless case, therefore, the transition is 'horizontal' as well as 'vertical', so that it is confined to a very small region of a potential energy curve or surface. The overlap in this region between the vibrational probability functions for starting and finishing states will then determine the efficiency of energy transfer for a fixed electronic transition probability. Figure 4.7 illustrates three possibilities: the curves given may be regarded as

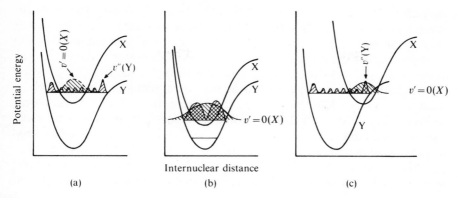

Fig. 4.7 The Franck–Condon principle in radiationless transitions. In (a) the geometries of states X and Y are similar, but the energy separations are large and the vibrational overlap is small. The separation is smaller in (b), and the overlap is larger, while (c) shows that significant overlap is possible for large energy separation if the geometries of X and Y are different.

potential energy curves for diatomic molecules, or as cross-sections through surfaces for more complex species. In Fig. 4.7(a), the two states, X and Y, are of similar geometry, but of widely different energy. The lowest vibrational level, $v' = 0$, in X has the same total energy as high v'' in Y. Because of the nature of the vibrational probability distributions, the overlap is small. In Fig. 4.7(b), however, the energy gap is much smaller, and the difference in vibrational quantum numbers v' and v'' is also smaller, with the result that there is far greater vibrational overlap. Thus, the efficiency of crossing will increase as $v'' \to 0$; i.e. crossing to a state will be favoured if the state can be populated near $v'' = 0$, which means that the *electronic* energy gap itself must be small. Only if the geometries of X and Y are different, as in Fig. 4.7(c), can there be rapid radiationless conversion between two states of widely separated electronic energies. In general, energy separations lie in the order $(S_1 - S_0) > (T_1 - S_0) > (S_1 - T_1)$, and the rate constants usually lie in the reverse order, even though the second two processes are formally spin-forbidden. We pointed out in Section 4.3 that molecular rigidity favours efficient fluorescence of a species. Without that rigidity, changes in geometry may occur, and $S_1 \rightsquigarrow S_0$ IC *can* depopulate S_1.

The importance of electronic energy separation in determining the efficiency of radiationless transfer is nicely illustrated by the effect of deuteration of some aromatic hydrocarbons on the rate of ISC ($T_1 \rightsquigarrow S_0$). Deuteration lowers C–H stretching frequencies, so that for a given electronic energy gap $\Delta E(T_1 - S_0)$, v'' in S_0 will be higher. Thus, in biphenyl, the rate constant, k_t, for the ISC process is $1.6 \times 10^{-1} \, \text{s}^{-1}$, but in decadeuterobiphenyl $k_t = 4.0 \times 10^{-2} \, s^{-1}$.   Again, $k_t = 4.0 \times 10^{-1} \, s^{-1}$ in naphthalene, but only $3.8 \times 10^{-2} \, s^{-1}$ in octadeuteronaphthalene. In neither case is the rate of radiative transition from T_1 to S_0 affected, so that the phosphorescence quantum yields are correspondingly higher in the deuterated compounds (e.g. 0.80 in $C_{10}D_8$ compared with 0.05 in $C_{10}H_8$).

The effect that the energy gap has on the rate of radiationless transition is illustrated in Fig. 4.8 for the $T_1 \rightsquigarrow S_0$ ISC in several aromatic hydrocarbons (in rigid solvents at 77 K). The electronic states involved are similar for each hydrocarbon, and the increase in rate of radiationless transition in going from benzene to anthracene follows the decrease in $\Delta E(T_1 - S_0)$. As a general rule, the larger the molecule, the smaller $\Delta E(T_1 - S_0)$, and ϕ_p is correspondingly low in large molecules. The differences in the photochemistry of species

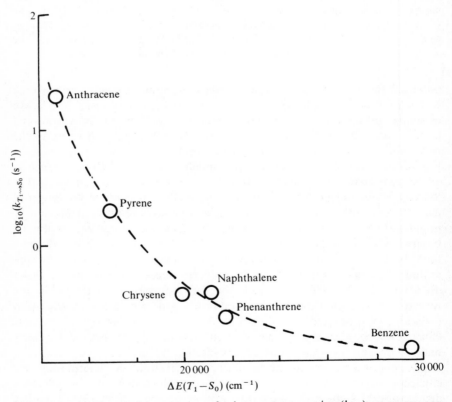

Fig. 4.8   Dependence of rate constant for intersystem crossing (k_{ISC}) on energy gap between triplet (T_1) and ground states (S_0) for several species. (From Birks, J. B. (1975). *Organic molecular photophysics*, Vol. 2, Tables 9.16 and 9.17. Wiley, New York.)

whose absorption system is predominantly $\pi - \pi^*$ or $n - \pi^*$ are at least in part a result of the increased likelihood of $S_1 \leadsto T_1$ ISC for $S_1 = {}^1(n, \pi^*)$.[†] Three factors work towards this increased likelihood of radiationless transition for ${}^1(n, \pi^*)$ states. First, both radiative and radiationless transitions to the S_0 ground state are partially forbidden, so that the ${}^1(n, \pi^*)$ state survives longer than a ${}^1(\pi, \pi^*)$ state and has more chance of undergoing ISC to T_1. Secondly, T_1 *may* be ${}^3(\pi, \pi^*)$, to which ISC from ${}^1(n, \pi^*)$ is favoured. Thirdly, the energy gap, $\Delta E(S_1 - T_1)$ is often small compared to the gap from ${}^1(\pi, \pi^*)$ states. The splitting for singlet to triplet (π, π^*) is often from about $10\,000$ cm^{-1} to $20\,000$ cm^{-1}, while that for (n, π^*) states is usually only 1500 cm^{-1} -5000 cm^{-1}. There are six possible arrangements of the energy levels where both (π, π^*) and (n, π^*) states exist, and Fig. 4.9 shows the types of species to which these arrangements apply. Note that in cases such as (a), where S_1 and T_1 are both (n, π^*) states, T_1 may be populated via the favourable ISC transition to ${}^3(\pi, \pi^*)$. In many carbonyl compounds phosphorescence is strong, and fluorescence weak or non-existent (typically $\phi_p/\phi_f > 1000$) because of this efficient crossing from S_1 to T_1; a fairly large fraction of the molecules reaching T_1 phosphoresce, the magnitude of the $T_1 \rightarrow S_0$ gap partly determining the rate of radiationless, $T_1 \leadsto S_0$ ISC, decay. The large $\Delta E(S_1 - T_1)$ and the unavailability of ${}^1(n, \pi^*) \rightarrow {}^3(\pi, \pi^*)$ transitions in hydrocarbons make $S_1 \leadsto T_1$ ISC less probable, and fluorescence becomes important as well as phosphorescence (and decay) from T_1. In connection with the importance of $\Delta E(S_1 - T_1)$, it is noteworthy that triphenylene, which has an abnormally low ΔE (6600 cm^{-1}) for a hydrocarbon, has $\phi_p/\phi_f = 5.9$ in a glass at 77 K, while for the more typical hydrocarbon, naphthalene: $\Delta E = 10\,900$ cm^{-1}; $\phi_p/\phi_f = 0.11$. In general, the separation $\Delta E(S_1 - T_1)$ decreases with increasing molecular size for $T_1 = {}^3(\pi, \pi^*)$, and ISC from S_1 to T_1 becomes more important. This result does *not* apply to the series benzene, naphthalene, anthracene.

Emitting triplet states have normally been characterized as (n, π^*) or (π, π^*) on the basis either of specific resolvable structure, e.g. of $C = O$ vibrations, or of phosphorescence lifetime measurements, the ${}^3(\pi, \pi^*)$ lifetimes being much greater than those of the ${}^3(n, \pi^*)$ state. The value of $\Delta E(S_1 - T_1)$ offers some indication of the natures of S_1 and T_1, although this kind of criterion must obviously be used with care. More recently, the influence of heavy atom environments (see later) on $T_1 \leftarrow S_0$ absorption intensities, or $T_1 \rightarrow S_0$ phosphorescence lifetimes, has been used to distinguish (n, π^*) and (π, π^*) states. Probabilities for transitions to or from ${}^3(\pi, \pi^*)$ states are increased by the heavy atom environment, and the technique has been used to show that, although T_1 for acetophenone ($C_6H_5COCH_3$) is ${}^3(n, \pi^*)$, for several substituted acetophenones the state is ${}^3(\pi, \pi^*)$.

[†] $\phi_{ISC} \sim 1$ at room temperature for $S_1 \leadsto T_1$ in several aromatic ketones (benzene solution); for hydrocarbons the efficiency is much less (e.g. $\phi_{ISC} \simeq 0.2$ for benzene).

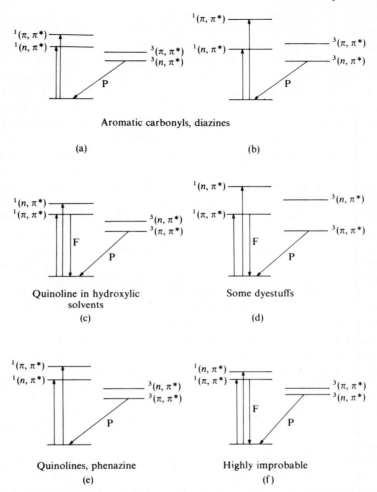

Fig. 4.9 The six possible arrangements of singlet and triplet (π, π^*) and (n, π^*) excited states; the diagram shows the types of molecule for which each arrangement is likely to apply. (From Wilkinson, F. and Horrocks, A. R. (1966). In *Luminescence in chemistry* (ed. by E. J. Bowen), p. 130. Van Nostrand, Princeton, NJ.)

The rate of spin-forbidden transitions may be perturbed by the external environment. Such an influence is seen in the effects of the addition of paramagnetic molecules to the solvent. Although O_2 and NO decrease phosphorescence yields because of their participation in efficient bimolecular quenching, they increase the rates both of optical transition and of ISC. Absorptions of the $T_1 \leftarrow S_0$ transition are also increased in intensity when the paramagnetic species is present, and, for example, the $T_1 \leftarrow S_0$ absorption in benzene ($\lambda \sim 310$–350 nm) practically disappears when the last traces of oxygen are removed. The most dramatic demonstration of the increase in $T \leftarrow S$ absorption is afforded by pyrene solutions, which are normally colourless but which turn deep red in the presence of high pressures of oxygen. Heavy atoms in an environment also increase the probability of $S \leftrightarrow T$ radiative and radiationless transitions by inducing appreciable spin–orbit coupling in the solute. Thus, solutions of anthracene and some of its derivatives become less fluorescent on addition of bromobenzene, while the triplet–triplet absorption intensity increases as a result of enhanced $S_1 \rightsquigarrow T_1$ ISC. As we noted earlier, these effects are most significant for transitions involving (π, π^*) excited states. Spin–orbit coupling is almost negligible in symmetric aromatic compounds, and it is here that perturbation by the environment has relatively the largest effect; significant spin–orbit coupling already exists in (n, π^*) states, and the effect of external perturbation is less pronounced. The effects are noticed in both solid and fluid solutions; for example, the phosphorescence lifetime of benzene in glasses at 4.2 K decreases from 16 s in CH_4 or Ar to 1 s in Kr and to 0.07 s in Xe: at the same time, the ratio ϕ_p/ϕ_f increases, and all processes $S_1 \overset{\text{ISC}}{\rightsquigarrow} T_1$, $T_1 \rightarrow S_0 + h\nu$ and $T_1 \overset{\text{ISC}}{\rightsquigarrow} S_0$ may proceed more rapidly in the solvents of higher atomic weight.

*Intra*molecular perturbation of transition probabilities is also important. Table 4.3 shows A_p, k_t, and ϕ_p/ϕ_f for naphthalene and some of its halo-derivatives. Substitution by one iodine atom increases the transition probability for optical emission (by a factor of nearly 10^4) and the rate of $T_1 \rightsquigarrow S_0$ ISC. Furthermore, the increase in ϕ_p/ϕ_f results mainly from increased probability of $S_1 \rightsquigarrow T_1$ ISC in the substituted molecules. Similar effects are observed on substitution in many other species. Indeed, the major effect of substitution on the photochemistry of a species seems to lie not so much in changes in energy levels (the first triplet and excited singlet levels lie at

Table 4.3. *Effect of halogen substitution in naphthalene on the rates of spin-forbidden processes* (*Data from Ermolaev, V. L. and Svitashev, K. K.* (*1965*). Opt. Spectrosc. 7, 399)

Compound	A_p (s^{-1}) $T_1 \rightarrow S_0 + h\nu$	k_t (s^{-1}) ISC $T_1 \rightsquigarrow S_0$	ϕ_p/ϕ_f
Naphthalene	0.05	0.39	0.09
1-Fluoronaphthalene	0.23	0.42	0.07
1-Chloronaphthalene	1.10	2.35	5.2
1-Bromonaphthalene	13.5	36.5	169
1-Iodonaphthalene	190	310	>760

21 300 cm^{-1}, 32 000 cm^{-1}, respectively, in naphthalene, and shift only to 20 500 cm^{-1}, 31 000 cm^{-1} in 1-iodonaphthalene), as in changes in the relative probabilities of fluorescence, phosphorescence, and the IC and ISC processes.

Strong intramolecular perturbations may also arise when certain metal ions are chelated to an organic molecule. It is interesting that the natural porphyrins chlorophyll and haemin display markedly different photo-chemical behaviour: chlorophyll has diamagnetic magnesium as its central ion, while haemin has paramagnetic iron (see Fig. 8.7 for the structure of chlorophyll).

This section is concluded with a description of a rather different kind of intramolecular energy transfer process in which energy is transferred from one *part* of a molecule to another (rather than between different *states*). Irradiation of the compound

with light that is absorbed solely by the naphthalene group leads to emission of anthracene fluorescence, and singlet excitation is transferred even though the absorbing and emitting groups are separated by an 'insulating chain' of three CH$_2$ groups. Triplet–triplet spatial energy transfer is also known. The naphthylalkyl benzophenones

absorb radiation at $\lambda \sim 366$ nm to excite exclusively the benzophenone-like unit to its singlet state; ISC to the triplet state follows, and excitation is then transferred to the naphthalene nucleus, which emits its characteristic phos-phorescence spectrum. The reaction is unlikely to proceed via singlet energy exchange followed by ISC in the naphthalene group, because S$_1$ for naphthal-ene is higher than that for benzophenone. Irradiation at $\lambda \sim 313$ nm excites S$_1$ of the naphthalene moiety, and singlet excitation can be transferred to the benzophenone unit, whence it eventually returns to the naphthalene as triplet excitation. Energy transfer of this kind can also occur in some rare-earth chelates. For example, irradiation of a low-temperature solution of europium benzoylacetonate leads to emission of the 613 nm Eu(^{5}D $\to$ ^{7}F) line. First, the β-diketone ligand absorbs light to reach a singlet state, and ISC then populates the triplet, which passes on its energy to the europium ion. Special interest in this process arises from the possibility of achieving population inversion and, hence, laser action (cf. Sections 2.3 and 5.7).

4.6 'Delayed' fluorescence

In the earliest experiments on the phosphorescence of fluorescein in boric acid glass, it was recognized that at least two mechanisms were operating to give long-lived emission: the processes were called α- and β-phosphorescence. β-Phosphorescence is the ordinary triplet–singlet emission described in previous sections, and its intensity is relatively insensitive to temperature. Several types of α-phosphorescence can be distinguished, and in this section we shall discuss the form that is sometimes known as E-type delayed fluorescence, after eosin, which displays the phenomenon (P-type delayed fluorescence, named after pyrene, will be mentioned in Section 5.5).

E-type delayed fluorescence spectra show features characteristic of the normal, short-lived, fluorescence. However, the emission decays at the same rate as phosphorescence; further, no emission is observed at low temperatures, and there is an activation energy for the process. This kind of delayed fluorescence arises from thermal activation of S_1 from $T_1(v = 0)$; the rate of activation is slow compared to the rate of loss of either T_1 or S_1, so that the decay of delayed fluorescence is determined by the decay of T_1. Figure 4.10 illustrates this excitation mechanism. The activation energy for the emission should be identical to $\Delta E(S_1 - T_1)$, and Table 4.4 shows the remarkable agreement obtained for some molecules between spectroscopic energy differences and experimental activation energies (calculated from the variation of emission intensity with temperature).

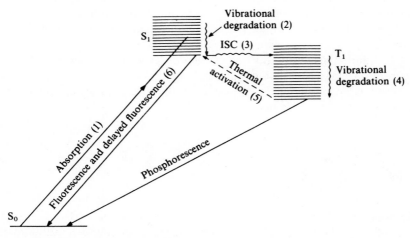

Fig. 4.10 Steps in the excitation mechanism for E-type delayed fluorescence.

Table 4.4 *Activation energies, E_a, for E-type delayed fluorescence and $\Delta E(S_1 - T_1)$ obtained spectroscopically (Data from Wilkinson, F. and Horrocks, A. R. (1968). In* Luminescence in chemistry (*ed. by E. J. Bowen), Table 7.7, p. 148, Van Nostrand, London*)

System	$\Delta E(S_1 - T_1)$ (kJ mol^{-1})	E_a (kJ mol^{-1})
Eosin in ethanol	45.6	40.5
Eosin in glycerol	43.9	42.6
Fluorescein in boric acid	38.5	33.4 ± 4.2
Proflavine in glycerol	33.4	33.4
Acriflavine in glucose	31.4 ± 2.1	33.4 ± 2.1
Acriflavine dimer in glucose	20.9 ± 2.1	23.0 ± 2.1
Acriflavine adsorbed on silica gel	20.9	33.4

4.7 Chemiluminescence

Some chemical reactions are accompanied by the emission of light, and the phenomenon is that of *chemiluminescence*. The excitation is *not* thermal; in flames (which show emission characteristic of, for example, the species C_2, CH, and OH), emission intensities may be higher than those expected from the flame temperature, and the radiation is chemiluminescent. Several natural chemiluminescent phenomena are well known, among them the light of glow-worms and fireflies, the glow of rotting fish, many bacteria, and the cold will-o'-the-wisp.

Detailed understanding of the excitation mechanism is restricted to few systems, mainly involving species in the gas phase. In this section we shall begin by mentioning some processes that are chemiluminescent in solution, and then select some gas-phase chemiluminescent reactions for further discussion.

Chemiluminescence in the firefly system is remarkably efficient, the overall quantum yield for emission approaching unity. The substrate molecule luciferin is oxidized in the presence of the energy-rich phosphate ATP (adenosine triphosphate: see Section 8.3).

$$\text{(4.27)}$$

An enzyme *luciferase* is necessary to trigger the process.

Artificial chemiluminescence from luminol (5-amino-2,3-dihydro-1,4-phthalazinedione) has been known from about the middle of the last century.

The reaction involves the oxidation of alkaline solutions of luminol, usually by H_2O_2 in the presence of the $Fe(CN)_6^{3-}$ ion,

$$\text{(4.28)}$$

and is one of the best-known chemiluminescent reactions. The green-blue emission derives from the aminophthalate ion. Many oxidation reactions that involve organic peroxides or hydroperoxides emit a narrow band at $\lambda = 634\,nm$: the same band is emitted from the reaction of sodium hypochlorite and H_2O_2. This band has been shown to be a 'dimol' emission involving *two* excited oxygen molecules in the $^1\Delta_g$ state. Radiation of the $^1\Delta_g \rightarrow {}^3\Sigma_g^-$ system is highly forbidden for electric dipole interactions, but emission is observed weakly, as a result of a magnetic dipole transition, at $\lambda = 1269\,nm$ (near infrared). The band at $\lambda = 634\,nm$ is equivalent to twice the energy of $O_2(^1\Delta_g)$ and arises when a collision complex $O_2(^1\Delta_g):O_2(^1\Delta_g)$ (a special kind of *excimer*, see Section 5.4) loses the whole of its excitation in a single quantum transition. If fluorescent substances (e.g. 9,10-substituted anthracenes) are present in the oxidation systems, then the characteristic fluorescence of these substances may be excited by intermolecular energy transfer (Chapter 5) from the energy-rich products of reaction.

The most efficient man-made chemiluminescent systems discovered so far are based on aromatic esters of oxalic acid (ethanedioic acid). Hydrogen peroxide decomposes the ester catalytically, and an energy-rich intermediate in the decomposition can transfer energy to a fluorescer present in the mixture. The ester CPPO (bis(carbopentoxy-3,5,6-trichlorophenyl) oxalate) is a typical example of the starting material.

$$\text{(4.29)}$$

Quantum yields reach values as high as 0.32, and the spectral characteristics can be selected by suitable choice of the fluorescer. Good fluorescers include 9,10-diphenylanthracene (blue), bis(phenylethynyl)anthracene (green), and rubrene (red). The commercial CYALUME® lightsticks use the oxalate system, the solution of oxalate ester and fluorescer being contained in an outer plastic tube, and the H_2O_2 in an inner breakable glass tube. Another remarkable chemiluminescent substance is tetrakis(dimethylamino)ethene, TKDE, which oxidizes spontaneously in air to produce a very bright and long-duration green chemiluminescence. The oxidation mechanism is very complex, and leads to excited molecules of the parent substance.

Both the oxalate and the TKDE systems have been proposed for use in emergency lighting systems, particularly in situations where the absence of heat or flame may be vital. Life-jacket markers using oxalate esters, and ropes incorporating TKDE that become illuminated when an evacuated covering is torn off, are amongst the many suggested applications. Potential artistic uses of chemiluminescence have been recognized, and necklaces containing re-packaged CYALUME® are a familiar sight. Chemiluminescence is of great value in analysis, partly because very low levels of emitted light may be detected (as low as a few photons per second). A standard biochemical assay for ATP uses nature's efficient chemiluminescent materials of firefly luciferin and luciferase, while TKDE is a sensitive indicator of oxygen, and the oxalate esters may be used to detect picomole (10^{-12} mole) quantities of hydrogen peroxide.

Two main types of process lead to chemiluminescence in the gas phase: recombination reactions and exchange reactions. Although chemiluminescent two-body recombination reactions are known, three-body recombination

$$A + B + M \xrightarrow{k_c} AB^* + M \tag{4.30}$$

(in which the third body M stabilizes the newly formed species AB; A + B is usually atom + atom, or atom + small molecule) is a more common source of intense chemiluminescence. If emission and quenching processes,

$$AB^* \xrightarrow{A_c} AB + h\nu \tag{4.31}$$

$$AB^* + M \xrightarrow{k_q} AB + M \tag{4.32}$$

are the only significant fates of AB^*, then the ordinary steady-state treatment yields a value for the intensity of chemiluminescence, I_c,

$$I_c = \frac{A_c k_c [A][B][M]}{A_c + k_q [M]} \tag{4.33}$$

If, as is often the case at moderately high pressures, $A_c \ll k_q[M]$, I_c is approximately independent of [M] and the process may *appear* to be second-order. The intensity will, however, depend on the *nature* of M, since both k_c and k_q vary with the chemical species.

Many of the investigations of gas-phase chemiluminescence of atom recombination reactions have been performed in flow systems using an electric discharge to produce relatively high concentrations of atoms (e.g. $1 - 10\%$). The atomic species are usually in their ground electronic state, and it is therefore not surprising that three-body chemiluminescence normally shows emission from levels *just* below the dissociation threshold for normal (unexcited) fragments. However, it is often found that the emission originates from an electronic state that does *not* correlate with (i.e. lie on the same potential curve or surface as) ground-state A and B: it seems that the emitting state is populated by radiationless transitions from a state that *does* correlate with the normal species. Thus, reaction (4.30) is a gross oversimplification of the excitation process, and several detailed mechanisms can be visualized that give overall third-order kinetic behaviour. Since one of the states with which normal A and B correlate may be the ground state of AB, a considerable fraction of AB might be formed unexcited and thus not give rise to chemiluminescence:

$$A + B + M \xrightarrow{k_c'} AB + M \qquad (4.30a)$$

One of the surprises about gas-phase chemiluminescence is the relatively large fraction of newly formed molecules that do, in fact, emit.

The familiar yellow afterglow of 'active' nitrogen (nitrogen that has been subjected to an electric discharge) is a result of chemiluminescent recombination of ground-state (^4S) nitrogen atoms. Most of the visible emission is of the 'First Positive' band system ($^3\Pi_g \to {}^3\Sigma_u^-$), so that reaction (4.30) must be written

$$N(^4S) + N(^4S) + M \to N_2(^3\Pi_g) + M \qquad (4.34)$$

even though two ^4S atoms do not correlate with the $^3\Pi_g$ state of N_2. Figure 4.11 shows potential energy curves for some states of N_2. It is thought that the $^3\Pi_g$ state is populated by a (collision-induced) radiationless transition from the $^3\Sigma_u^+$ state. Measurements of the absolute emission efficiency show that between one-third and one half of the total recombination proceeds via reaction (4.34) to the $^3\Pi_g$ state. If the rate of formation of $^3\Sigma_u^+$ and ground, $^1\Sigma_g^+$, states is in the statistically expected ratio of three to one, the crossing $^3\Sigma_u^+ \to {}^3\Pi_g$ must be highly efficient (about 66%) at the pressures employed.

The kinetic behaviour of the First Positive band emission is well represented by eqn (4.33). At pressures above about 1 mmHg the intensity is proportional to $[N]^2$, and independent of the concentration of M but not of

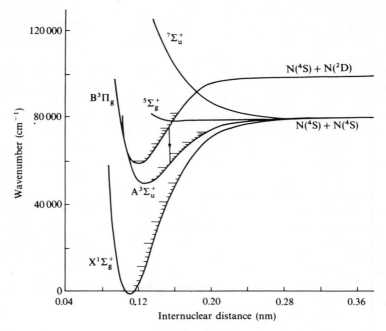

Fig. 4.11   Potential energy curves for some states of N_2. (From Thrush, B. A. (1966). *Chem. Br.* **2**, 287.)

its nature; at lower pressures the intensity becomes proportional to $[N]^2[M]$.

Another highly efficient chemiluminescent recombination is the reaction

$$O + NO + M \rightarrow NO_2^* + M \qquad (4.35)$$

which gives rise to the so-called 'air afterglow'. The intensity is proportional to $[O][NO]$, and is dependent on the nature of the third body, at ordinary pressures; as predicted by eqn (4.33), the intensity becomes proportional to $[M]$ at sufficiently low pressures (around 5×10^{-2} mmHg). Bands in the emission spectrum show that the air afterglow involves the same electronic transition of NO_2 as that responsible for the visible absorption; the short wavelength limit of the chemiluminescence corresponds exactly to the onset of predissociation in the absorption spectrum.

Many other recombination reactions are chemiluminescent in the gas phase. The carbon monoxide flame emission bands are a result of $O + CO$ recombination, and the spectrum shows features due to transition from the bent $^1B_2$ state of CO_2 to the linear $^1\Sigma_g^+$ ground state. Excitation of atomic spectra in some flames containing metal or metal salt vapours often involves a chemiluminescent recombination of a different kind. The overall excitaion process can be represented by

$$H + X + M \rightarrow HX + M^* \tag{4.36}$$

where X = H or OH and M is the metal atom. Other processes in which chemiluminescence has been studied are $N + O$, $O + O$, $H + NO$, $O + SO$, and various halogen atom recombinations.

The photochemistry we have discussed so far in this book has always involved electronically excited states. A number of exothermic transfer reactions of the general type

$$A + BC \rightarrow AB^v + C \tag{4.37}$$

yield a product molecule AB with a high degree of vibrational excitation in the newly formed bond (the superscript v is used to represent this excitation). This vibrational excitation does not correspond to the kinetic temperature of the gas, so that, if the vibration is 'infrared active' (i.e. has associated with it an oscillating dipole), chemiluminescent emission may be observed in the red or near-infrared regions of the spectrum. The reaction of atomic hydrogen with halogens is typical of the reactions that lead to infrared chemiluminescence,

$$H + X_2 \rightarrow HX^v + X \tag{4.38}$$

With $X_2 = Cl_2$, vibration–rotation emission spectra for transitions of $\Delta v = 0$, 1, and 2 have been observed. At pressures low enough to avoid collisional degradation, up to six vibrational quanta are present in the newly formed HCl. The vibrational energy bears no relation to an equilibrium distribution, although the fine structure of the spectrum suggests that rotational energy is in thermal equilibrium at a temperature not more than 100°C above that of the reaction vessel. The exothermicity of the reaction is rather less than the energy of the observed six vibrational quanta; however, the total energy available for excitation is that from the *top of the activation barrier* to the product HCl (i.e. equivalent to $E_a + |\Delta H|$).

Vibrational chemiluminescence from exchange reactions currently arouses great interest because of its application in studies of theoretical kinetics. The spectroscopic structure of vibrational chemiluminescence can be compared with the vibrational and rotational excitation predicted by computer simulation of reaction trajectories on the potential energy surface linking reactants and products. Such studies provide information both on the nature of the potential energy surface and on the detailed dynamics of reactive interactions (cf. Section 7.6).

Several four-atom

$$A + BCD \rightarrow AB^v + CD \tag{4.39}$$

reactions display infrared chemiluminescence. An example of such a process is the highly exothermic (335kJ mol^{-1}) reaction

$$H + O_3 \rightarrow OH^v + O_2 \tag{4.40}$$

The observed OH emission extends from the infrared into the long-wavelength end of the visible spectrum, and the bands are forbidden 'overtone' vibrational transitions ($\Delta v = 4$ or 5). The overtone bands of OH are observed in the glow of the night sky, and reaction (4.40) is believed to be the source of OH excitation in the upper atmosphere.

We have seen in this section that species excited in chemical reactions take part in exactly the same processes (emission, quenching, intramolecular energy transfer) as those formed by absorption of light. It is hoped that the discussion of chemiluminescent phenomena has also shown the interrelation between reaction kinetics, spectroscopy and photochemistry.

Bibliography

Parker, C. A. (1969). *Luminescence of solutions.* Elsevier, Amsterdam.
Bowen, E. J. (ed.) (1968). *Luminescence in chemistry.* Van Nostrand, Princeton, NJ.
Calvert, J. G. and Pitts, J. N., Jr (1966). *Photochemistry,* Chapter 4. John Wiley, Chichester and New York.
Barltrop, J. A. and Coyle, J. D. (1978). *Principles of photochemistry,* Chapters 3 and 4. John Wiley, Chichester and New York.
Birks, J. B. (1973). *Organic molecular photophysics,* Vol. 1. John Wiley, Chichester and New York.
Birks, J. B. (1975). *Organic molecular photophysics.* Vol. 2. John Wiley, Chichester and New York.
Becker, R. S. (1969). *Theory and interpretation of fluorescence and phosphorescence.* John Wiley, Chichester and New York.
Parker, C. A. (1966). Photoluminescence as an analytical technique. *Chem. Br.* **2**, 249.
Lim, E. (ed.) (1969). *Molecular luminescence.* W. A. Benjamin, New York.
Stockburger, M. (1973). Fluorescence of aromatic molecular vapours. In *Organic molecular photophysics* (ed. J. B. Birks) Vol. 1. John Wiley, Chichester and New York.
Birks, J. B. (1975). Photophysics of aromatic molecules—a postscript. In *Organic molecular photophysics* (ed. J. B. Birks) Vol. 2. John Wiley, Chichester and New York.
Robin M. B. (1974, 1975, 1985). *Higher excited states of polyatomic molecules.* Vols 1, 2 and 3. Academic Press, New York.

Special topics
Sections 4.4 and 4.5

McGlynn, S. P., Azumi, T., and Kinoshita, M. (1969). *Molecular spectroscopy of the triplet state.* Prentice-Hall, Englewood Cliffs, NJ.
Wild, U. P. (1975). Characterization of triplet states by optical spectroscopy. *Top. Curr. Chem.* **55**, 1.

Pratt, D. W. (1979). Magnetic properties of triplet states. In *Excited states* (ed. E. C. Lim) Vol. 4. Academic Press, New York.

Lin, S. H. (ed.) (1980). *Radiationless transitions*. Academic Press, New York.

Freed, K. F. (ed.) (1978). Radiationless transitions in molecules. *Acc. Chem. Res.* **11**, 74.

Jortner, J., Rice S. A., and Hochstrasser, R. M. (1969). Radiationless transitions in photochemistry. *Adv. Photochem.* **7**, 149.

Formosinho, S. J. (1986). Radiationless transitions and photochemical reactivity. *Pure Appl. Chem.* **58**, 1173.

Wilkinson, F. (1975). Triplet quantum yields and singlet–triplet intersystem crossing. In *Organic molecular photophysics* (ed. J. B. Birks) Vol. 2. John Wiley, Chichester and New York.

Henry, B. R. and Siebrand, W. (1973). Radiationless transitions. In *Organic molecular photophysics* (ed. J. B. Birks) Vol. 1. John Wiley, Chichester and New York.

Koziar, J. C. and Cowan, D. O. (1978). Photochemical heavy atom effects. *Acc. Chem. Res.* **11**, 334.

Sections 4.4 and 4.6

Parker, C. A. (1964). Phosphorescence and delayed fluorescence from solution. *Adv. Photochem.* **2**, 305.

Section 4.7

DeLuca, M. and McElroy, W. D. (eds.) (1981). *Bioluminescence and chemiluminescence: Basic chemistry and analytical applications*. Academic Press, New York.

Cormier, M. J., Hercules, D. M., and Lee, J. (eds.) (1973). *Chemiluminescence and bioluminescence*. Plenum Press, New York.

Schuster, G. B. and Schmidt, S. P. (1982). Chemiluminescence of organic compounds. *Adv. Phys. Org. Chem.* **18**, 187.

Adam, W. J. (1975). Biological light. *J. Chem. Educ.* **52**, 138.

Gill, S. K. (1983). New developments in chemiluminescence research. *Aldrichimica Acta* **16**, 59.

Smith, I. W. M. (1975). The production of excited species in simple chemical reactions. *Adv. Chem. Phys.* **28**, 1.

Toby, S. (1984). Chemiluminescence in the reactions of ozone. *Chem. Rev.* **84**, 277.

Polanyi, J. C. (1966). Energy distribution among reaction products and infrared chemiluminescence. *Chem. Br.* **2**, 151.

Thrush, B. A. (1966). Formation of electronically excited molecules in simple gas reactions. *Chem. Br.* **2**, 287.

Rauhut, M. M. (1969). Chemiluminescence from concerted peroxide decomposition reactions. *Acc. Chem. Res.* **2**, 80.

Gunderman, K. (1974). Recent advances in research on the chemiluminescence of organic compounds. *Top. Curr. Chem.* **46**, 61.

5

Energy transfer: emission processes (2)

5.1 Intermolecular energy transfer

The *intra*molecular exchange of electronic energy between different states, and its exchange with vibrational energy, play an important part in determining the photochemical behaviour of a molecule (Chapter 4). *Inter*molecular exchange of energy between two discrete partners may also take place: the 'acceptor' (A), which receives excitation from the 'donor' (D), then participates in those processes open to it as an electronically excited species. *Photosensitized* phenomena, in which the change of interest occurs in a species other than the one that absorbed radiation, are believed to be of great significance in photobiology; they also provide valuable insights into photophysical processes.

Franck predicted in 1922 that electronic excitation could be exchanged between atoms, and Cario and Franck subsequently demonstrated *sensitized fluorescence* in a mixture of mercury and thallium vapours. The mixture was irradiated with the $\lambda = 253.7$ nm resonance line of mercury, to which thallium vapour is transparent; emission was observed from the thallium. Absorption of light by Hg raises it to the resonance level, $^3P_1$, and energy is then transferred to the thallium.

$$Hg + h\nu_{\lambda = 253.7 \text{ nm}} \rightarrow Hg^* \tag{5.1}$$

$$Hg^* + Tl \rightarrow Hg + Tl^* \tag{5.2}$$

$$Tl^* \rightarrow Tl + h\nu \tag{5.3}$$

Exchange of energy between two different species is not so restricted with regard to exact equivalence of internal energy between initial and final states as in the case of intramolecular exchange, since an energy excess can be taken up by translation (or, more rarely, a deficiency supplied by the kinetic energy of collision). Ten different types of energy exchange can be classified according to the modes (electronic, vibration, rotation, and translation) between which the exchange occurs; except in the rare case of exact energy resonance, some energy is always converted to or from translation. We have already given implicit consideration to the degradation of electronic excitation to vibration, rotation or translation, in our discussion of physical quenching of fluorescence. Much of the following discussion will, however, be directed towards electronic–electronic energy exchange, and it will be assumed that *excess* energy goes into other modes of excitation. Where the absorption spectrum of the acceptor overlaps the emission spectrum of the donor, there are, of course, quantized vibronic levels of A and D for which energy exchange is isoenergetic, and no increase in kinetic energy is needed. For electronic energy exchange between atoms, some translational energy is almost always released.

Several different mechanisms of electronic energy transfer are believed to operate under different circumstances. The first of these is the so-called 'trivial' mechanism of radiative transfer, which can be represented by the processes

$$D^* \rightarrow D + h\nu \tag{5.4}$$

$$A + h\nu \rightarrow A^* \tag{5.5}$$

The mechanism is trivial in name and simplicity only, since it is the one energy transfer mechanism that can operate over very large separation of D and A: the interaction necessarily follows the laws of light propagation. Radiative energy transfer is all-important to our existence, because it is how we receive the energy of reactions occurring in the Sun; and the related radiative energy transfer processes occurring in upper and lower atmospheres establish the temperature equilibria and meteorological conditions upon which we depend. The efficiency of radiative transfer is a function of the overlap between the emission spectrum of D and the absorption spectrum of A (a factor that appears in all transfer mechanisms), and also of the size and shape of the sample: since D^* will emit in all directions, the probability of radiative transfer increases with sample volume. It will be obvious that experiments designed to study non-radiative energy transfer must eliminate or make due allowance for the radiative process.

'Short-range' energy transfer arising from *exchange interaction* occurs over intermolecular or interatomic distances (henceforth referred to as r) not much exceeding the collision diameter; the interaction decreases in a complex fashion with r raised to a high power.

'Long-range' transfer may arise by sequential short-range excitation of many species so that the excitation appears ultimately at a place 'distant' from the original location of excitation. Long-range transfer is, however, predicted to occur also by a direct mechanism involving electrical, or coulombic, interactions between transition dipoles (or higher multipoles). These multipoles are the ones involved in optical interactions with the electric vector of radiation: the usual optical selection rules apply to both the transitions $D^* \rightarrow D$ and $A \rightarrow A^*$, and dipole–dipole (dd) interactions are stronger than dipole–quadrupole (dq) interactions, and so on. For a dd interaction, theory predicts that the strength of the interaction should fall off as $1/r^6$, and relatively long-range energy exchange becomes possible.

An even longer-range transfer, showing a $1/r^3$ dependence, may occur in crystals, solid solutions, and some fluids, as a result of *exciton* migration. The concept of the exciton was introduced by Frenckel to interpret certain crystal spectra; an electron–hole pair was looked upon as an entity that could move about the crystal as a result of interactions between lattice sites. For our purposes, we can regard the electronic excitation in an irradiated species as an exciton that is free to wander over a considerable number of lattice sites. We shall not discuss this mechanism further in this book.

Energy may be transferred from one excited species, D*, to another, A*, already possessing some excitation, to raise the latter to a higher electronic state A**,

$$D^* + A^* \rightarrow D + A^{**} \tag{5.6}$$

This process of *energy-pooling* must be relatively slow at low concentrations of D* and A*, since bimolecular collision of the excited species will be a rare event. Formation of excited species by energy-pooling has, however, been recognized in several systems, and provides the mechanism for *P-type delayed fluorescence*. Energy-pooling may also permit the occurrence of chemical reactions that require more energy than is available in a single quantum of radiation, and such energy storage may be a necessary step in several photobiological systems.

5.2   Short-range, collisional, energy transfer

Energy transfer by exchange interaction may be thought of as a special kind of chemical reaction in which the chemical identity of the partners A and D does not change, but in which excitation is transferred from one to the other. The transition state is then expected to possess a separation between A and D not greatly different from the sum of the gas-kinetic collision radii, and energy transfer by the exchange mechanism is probably important only for values of r of this order. In common with other chemical processes, energy transfer can be efficient only if the potential energies of reactants and products are connected by a *continuous* surface that describes the potential energy of the system as a function of the several interatomic distances: a reaction occurring 'on' such a surface is said to proceed *adiabatically*. In other words, the reactants and products must correlate with each other and with the transition state. Most chemical reactions involving ground-state partners can occur adiabatically, but in processes such as energy exchange, where several electronic states are involved, the requirement for adiabatic reaction may impose some restrictions on the possible states for A, A* and D, D* if there is to be efficient

Table 5.1 *Multiplicities of transition states for given multiplicities of two separated species*

Separated species	Transition state
singlet + singlet	singlet
singlet + doublet	doublet
singlet + triplet	triplet
doublet + doublet	singlet, triplet
doublet + triplet	doublet, quartet
triplet + triplet	singlet, triplet, quintet

transfer of excitation. In atoms or small molecules there must be correlation of electron spin, orbital momentum, parity, and so on. However, correlation in complex molecules of low symmetry usually only involves the electron spin. To test for correlation, the possible total spin of the transition state is calculated from the individual spins of the reactants by the addition of the quantum vectors (see Section 2.5 for addition of quantized vectors in single atoms or molecules). Thus, for reactants A and B with spins S_A and S_B, the total spin of the transition state can take magnitudes $|S_A + S_B|$, $|S_A + S_B - 1|, \ldots, |S_A - S_B|$. It is then necessary to see whether the products, X and Y, can also give at least one of the same total spin magnitudes in the transition state. Table 5.1 shows multiplicities in the transition state that can arise from some multiplicities of two separated species. Thus we can see, by use of the table, that all of the processes

$$
\begin{aligned}
&\qquad\qquad\text{[singlet]}\leftrightarrow C(\text{singlet}) + D(\text{singlet}) \\
&A(\text{triplet}) + B(\text{triplet})\leftrightarrow[\text{singlet or triplet}]\leftrightarrow C(\text{doublet}) + D(\text{doublet}) \qquad (5.7) \\
&\qquad\qquad\text{[singlet or triplet]}\leftrightarrow C(\text{triplet}) + D(\text{triplet})
\end{aligned}
$$

can occur adiabatically. On the other hand, the reaction

$$A(\text{singlet}) + B(\text{triplet}) \not\leftrightarrow [\text{no common multiplicity}] \not\leftrightarrow C(\text{singlet}) + D(\text{singlet}) \tag{5.8}$$

cannot proceed (in either direction) adiabatically. Arguments of this kind were used by Wigner to derive the *Wigner Spin Correlation Rules* (e. g. triplet + triplet → triplet + triplet, etc.).

Although a reaction is likely to be efficient only if it is adiabatic, non-adiabatic reactions can also occur.[†] We may look upon a non-adiabatic

---

[†] The following discussion is an oversimplification, but adequate for the present purposes. Reactions that are, by common consent, called non-adiabatic—such as those that are spin-forbidden—may in fact occur adiabatically but with low efficiency.

reaction as one in which crossing occurs between two intersecting or closely approaching potential energy surfaces. The crossing process is governed by the ordinary selection rules for radiationless transitions. In particular, a spin-forbidden reaction cannot proceed adiabatically because no common spin states can be written for the transition complex, and potential surfaces for transition states derived from reactants and products must be of different multiplicity. Hence, $\Delta S \neq 0$ for the *intramolecular* energy transfer, the crossing is of low probability (cf. Section 4.5), and the efficiency of the non-adiabatic *intermolecular* energy transfer is small.

Energy exchange in molecules occurs efficiently if the amount of kinetic (translational) energy that must be liberated is small; thus we may expect rapid exchange of energy between vibronic levels in near-resonance. Excess vibrational energy will be degraded rapidly (at least in condensed-phase systems), and the acceptor molecule left in its ground vibrational state. The *apparent* energy gap is therefore the electronic difference between D* and A*, although this is *not* ΔE for the actual transfer process. Anomalies in the dependence on ΔE of rate of transfer may arise because of the difference between actual and apparent energy gaps. Table 5.2 shows measured rate constants for several triplet–triplet energy exchange processes

$$D^*(T_1) + A(S_0) \rightarrow D(S_0) + A^*(T_1) \tag{5.9}$$

and lists the energy difference between $D^*(T_1, v = 0)$ and $A^*(T_1, v = 0)$. A negative ΔE in the table represents an exothermic reaction. So long as the reaction is exothermic, energy transfer is fast (it approaches the diffusion-controlled limit: cf. pp. 72 and 106), but when the reaction becomes endothermic, the rate rapidly drops off.

Table 5.2 *Rate constants for triplet energy exchange (Data quoted by Turro, N. J. (1978). Modern molecular photochemistry, Table 9.3. Benjamin/Cummings, Menlo Park, CA) (Solvents n-hexane, iso-octane, or benzene)*

Donor	Acceptor	ΔE (cm^{-1})	Rate constant (dm^3 mol^{-1} s^{-1})
Acetophenone	Naphthalene	-4200	1×10^{10}
Benzophenone	Naphthalene	-2800	1×10^{10}
Triphenylene	Naphthalene	-2100	2×10^9
Naphthalene	Biacetyl	-1700	9×10^9
Bromonaphthalene	Biacetyl	-1000	3×10^9
Biacetyl	Bromonaphthalene	1000	3×10^7
Biacetyl	Naphthalene	1700	2×10^6
Naphthalene	Triphenylene	2100	$< \quad 10^4$
Naphthalene	Benzophenone	2800	$< \quad 10^4$

The efficiency of energy transfer occurring by the exchange interaction mechanism is related to whether the process can take place adiabatically, but *not* to whether optical selection rules permit radiative transitions in donor and acceptor; this behaviour is one way in which exchange interaction and long-range coulombic interaction may be distinguished. For example, in the exchange interaction excitation of triplet states by triplet benzophenone, the efficiency of energy transfer is roughly the same both for naphthalene and for 1-iodonaphthalene. We saw in the last chapter (p. 89) that the $T_1 \rightarrow S_0$ radiative transition is by a factor of at least 1000 more probable in the substituted molecule, so that in this case the optical transition probability in the naphthalene molecule does not seem to affect the probability of energy transfer to it.

Energy transfer and emission from an excited donor are competitive, and kinetic analysis of sensitized emission phenomena can yield information about rate constants for the transfer process. A simple excitation scheme (in which A does not absorb) can be written for sensitized emission:

		rate:	
$D + h\nu + D^*$	absorption	I_{abs}	(5.10)
$D^* \rightarrow D + h\nu_D$	donor emission	$A_D[D^*]$	(5.11)
$D^*[+M] \rightarrow D[+M]$	donor energy loss	$k_D^q[D^*]$	(5.12)
$D^* + A \rightarrow D + A^*$	energy transfer	$k_e[A][D^*]$	(5.13)
$A^* \rightarrow A + h\nu_A$	acceptor emission	$A_A[A^*]$	(5.14)
$A^*[+M] \rightarrow A[+M]$	acceptor energy loss	$k_A^q[A^*]$	(5.15)

Reactions (5.12) and (5.15) include all non-radiative loss processes for D^* and A^*, respectively (e.g. radiationless intramolecular energy degradation and bimolecular quenching). For constant [M] these processes may be described by a single pseudo-first-order rate constant k^q. Solution of the stationary-state equations for $[D^*]$ and $[A^*]$ yields the results

$$I_D = \frac{A_D I_{abs}}{A_D + k_D^q + k_e[A]} \tag{5.16}$$

and

$$I_A = \frac{A_A k_e[A] I_{abs}}{(A_A + k_A^q)(A_D + k_D^q + k_e[A])} = \frac{A_A}{A_D} \cdot \frac{k_e[A]}{A_A + k_A^q} \cdot I_D \tag{5.17}$$

where I_D, I_A are the intensities of emission from donor and acceptor, respectively. Thus, if I_D^0 is the emission intensity from the donor when $[A] = 0$, then

$$\frac{I_D^0}{I_D} = 1 + \frac{k_e}{A_D + k_D^q}[A] \qquad (5.18)$$

That is, the quenching by the acceptor of donor emission follows a Stern–Volmer law, and values for k_e may be calculated if $(A_D + k_D^q)$ is known. Figure 5.1 shows some results for the quenching by biacetyl of fluorescence from various donors. From the slope of the graph and the known value of $(A_D + k_D^q)$ it may be shown, for example, that $k_e = 3.7 \times 10^{10}$ dm^3 mol^{-1} s^{-1}. For transfer from toluene a further check on k_e is available, via eqn (5.17), from the ratio of acceptor to donor intensities, I_A/I_D, if $(A_A + k_A^q)$ is known. If, as may be the case, A also absorbs the exciting radiation, eqn (5.17) must be modified; eqns (5.16) and (5.18) remain unaltered. Table 5.3 shows rate constants for singlet–singlet energy transfer from several donors to biacetyl in hexane solution.

The large rate constants and their relative insensitivity to the nature of the donor together suggest that the transfer of energy is diffusion-controlled. The Debye equation for the rate constant in a diffusion-controlled reaction of species of similar size (eqn 4.8) gives a value of $k_e \sim 2.4 \times 10^{10}$ dm^3 mol^{-1} s^{-1} for hexane at 28°C, in qualitative agreement with the data of Table 5.3. Even better agreement is obtained if the equation for the diffusion-controlled rate constant is modified for the case where there is no friction between partners:

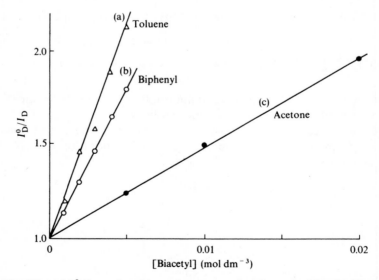

Fig. 5.1 Plot of I_D^0/I_D against biacetyl concentration for quenching by biacetyl of fluorescence from (a) toluene, (b) biphenyl, and (c) acetone. (From Wilkinson, F. and Dubois, J. T. (1963). *J. Chem. Phys.* **39**, 377.)

Table 5.3 *Rate constants for energy transfer to biacetyl (hexane solution at 28°C)* (*Data of Wilkinson, F. and Dubois, J. T.* (*1963*) J. Chem. Phys. *39, 377.; and Dubois, J. T. and Van Hemert, R. L.* (*1964*) J. Chem. Phys. *40, 923*)

Donor	$k_e(\mathrm{dm^3\ mol^{-1}\ s^{-1}})$
Benzene	3.3×10^{10}
Toluene	3.7×10^{10}
o-Xylene	3.5×10^{10}
m-Xylene	3.3×10^{10}
p-Xylene	3.4×10^{10}
Pentamethylbenzene	4.5×10^{10}
Hexamethylbenzene	4.0×10^{10}
Ethylbenzene	3.6×10^{10}
n-Propylbenzene	4.2×10^{10}
n-Butylbenzene	3.4×10^{10}
Naphthalene	2.2×10^{10}

the value obtained is $k_e \sim 3.5 \times 10^{10}\ \mathrm{dm^3\ mol^{-1}\ s^{-1}}$ in hexane (28°C). The rate constants for *triplet* energy transfer shown in Table 5.2 also approach the diffusion-controlled value when ΔE is negative.

Triplet–triplet energy transfer is, in fact, sometimes treated as though it were a different phenomenon from singlet–singlet transfer. However, so far as the exchange interaction mechanism is concerned, the fact that both A and D change their spin multiplicity is of no account, since the reaction is adiabatic. Observed differences in photochemical behaviour arise from the long radiative lifetimes of triplet states. In media in which quenching and radiationless energy degradation are slow (e.g. in rigid glasses), the long actual lifetime of the triplet donor means that even inefficient energy transfer can compete successfully with other loss processes. At the same time, sensitized phosphorescence is likely to be seen only in systems where radiationless loss and quenching are not the major fates of the triplet acceptor (e.g. again in rigid glasses, or with acceptors such as biacetyl).

Absorption of the exciting radiation by both donor and acceptor can complicate interpretation of sensitized fluorescence in studies of singlet–singlet transfer. Triplet–triplet exchange, on the other hand, can be investigated in systems where only the donor absorbs. Suitable choice of donor and acceptor permits the triplet of D to lie *above* that of A, so that $D^* \rightarrow A$ transfer can occur, at the same time that $S_1(D)$ is below $S_1(A)$, so that excitation of $S_1(D)$ can occur at wavelengths longer than those that could excite A directly. The required order of energy levels is often found for aromatic carbonyl compounds as donors, and aromatic hydrocarbons as acceptors. Figure 5.2 shows the energies of the triplet and singlet states in

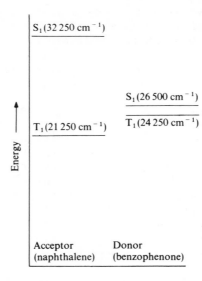

Fig. 5.2 Energy levels of excited singlet and triplet states in naphthalene and in benzophenone.

benzophenone and naphthalene. Irradiation at $\lambda = 366$ nm of a benzophenone–naphthalene mixture (in a rigid glass at $-180°C$) leads to emission from naphthalene, and the spectrum is identical with that of the phosphorescence of naphthalene. Light at $\lambda = 366$ nm is, however, absorbed only by the benzophenone, and the emission from naphthalene is *sensitized* phosphorescence. The excitation scheme is

$$B(S_0) + h\nu_{\lambda \sim 366 \text{ nm}} \rightarrow B(S_1) \qquad (5.19)$$

$$B(S_1) \overset{\text{ISC}}{\leadsto} B(T_1) \qquad (5.20)$$

$$B(T_1) + N(S_0) \rightarrow B(S_0) + N(T_1) \qquad (5.21)$$

$$N(T_1) \rightarrow N(S_0) + h\nu_{\text{phos}} \qquad (5.22)$$

Direct identification by ESR of the triplet state of naphthalene has proved possible in similar glasses of benzophenone–naphthalene mixtures irradiated at $\lambda = 366$ nm.

Much of our discussion about energy transfer processes has suggested that population of acceptor levels *above* those of the donor (i.e. with ΔE positive) can occur only with an activation energy for reaction equal to ΔE. Indeed, triplet–triplet energy transfer in solution was first demonstrated, albeit circumstantially, by the quenching of biacetyl phosphorescence only by those quenchers whose triplet level lay below that of biacetyl. The implication is that quenching involves triplet–triplet energy exchange, and subsequent experiments have, in fact, detected quencher triplets by their absorption spectra. There are, however, some most interesting cases where, although energy transfer is of reduced efficiency, the activation energy is much smaller than the endothermicity. For example, the rate of transfer to *cis*-stilbene (*cis*-1,2-diphenylethene) of triplet excitation drops by a factor of rather less than two as the donor energy drops from approximately the same as that of the *cis*-stilbene triplet (~ 240 kJ mol^{-1}) to 13 kJ mol^{-1} less. The rate for the reaction, endothermic by 13 kJ mol^{-1}, should be slower by a factor of nearly 150 than the thermoneutral process on the basis of activation energies. It appears that energy released on modification of the molecular geometry in going from ground to triplet states of the acceptor can contribute towards the total excitation energy: the time scale for bimolecular reaction is quite sufficient to allow accompanying skeletal motion. Acceptor molecules having large separations between the (0, 0) bands in emission and absorption are therefore most likely to exhibit this behaviour, a view confirmed by such experimental evidence as there is. Where geometrical rearrangement cannot provide the necessary energy, the rate of transfer drops off rapidly as the process becomes more endothermic. With *cis*-stilbene as acceptor, the rate begins to drop off rapidly for ΔE greater than about 20 kJ mol^{-1}, which suggests that this is the maximum amount of energy available from geometrical rearrangement.

5.3 Long-range, coulombic, interactions

For many energy transfer processes, the interaction takes place when the partners are separated by more than the sum of the gas–kinetic collision radii. For example, energy transfer between excited singlet states of hydrocarbons occurs as fast as spontaneous decay at concentrations in benzene around $10^{-2}-10^{-3}$ mol dm^{-3}, which corresponds to a distance, r, between exchanging molecules of about 5 nm, or about 10 times the collision diameter. Again, the measured rate constants for transfer of excitation in the hydrocarbons seem greatly to exceed the diffusion limited rate, and do not depend on solvent viscosity. Thus, rate constants for the process

$$1\text{-chloroanthracene}(S_1) + \text{perylene}(S_0) \rightarrow 1\text{-chloroanthracene}(S_0)$$
$$+ \text{perylene}(S_1) \qquad (5.23)$$

are about 1.5×10^{11} dm^3 mol^{-1} s^{-1} in room temperature solutions in benzene or in liquid paraffin, and virtually the same in a glass at $-183°C$.

Förster has derived an equation for the rate per second, k_e, of energy transfer by the dipole–dipole *inductive resonance*, or coulombic interaction,

Energy transfer: emission processes (2)

mechanism. The equation reduces to the approximate form

$$k_e \sim 1.25 \times 10^{-25} \frac{\phi_D^f}{n^4 \tau_D r^6} \int_0^\infty f_D(\bar{\nu}) \varepsilon_A(\bar{\nu}) \frac{d\bar{\nu}}{\bar{\nu}^4} \qquad (5.24)$$

where r is in cm, ϕ_D^f is the fluorescence quantum yield for the *donor*, $f_D(\bar{\nu})$ the normalized spectral distribution (in quantum units) of the emission and τ_D its mean lifetime, $\varepsilon_A(\bar{\nu})$ is the molar decadic extinction coefficient of the *acceptor* at a frequency $\bar{\nu}$ in cm^{-1}, and n is the solvent refractive index.

We may define a 'critical distance', r_0, as that intermolecular separation at which energy transfer and spontaneous decay are equal, i.e. when $k_e = 1/\tau_D$. From eqn (5.24) we calculate r_0 to be given by

$$r_0 \sim \left[1.25 \times 10^{-25} \frac{\phi_D^f}{n^4} \int_0^\infty f_D(\bar{\nu}) \varepsilon_A(\bar{\nu}) \frac{d\bar{\nu}}{\bar{\nu}^4} \right]^{1/6} \qquad (5.25)$$

Table 5.4 compares values of k_e and r_0 calculated from eqns (5.24) and (5.25) with some experimental values. The agreement between observed and predicted values of k_e is fair; theory and experiment concur well in deriving values for the critical distances, which are considerably in excess of the kinetic collision radii.

Table 5.4 *Comparison of experimentally determined exchange rate constants and critical distances with those calculated from eqns (5.24) and (5.25) (Data from Wilkinson, F. (1964). Adv. Photochem. 3, 241, Table III)*

Donor	Acceptor	$10^{-10} k_e$ (dm^3 mol^{-1} s^{-1})		r_0 (nm)	
		Theory	*Experiment*	*Theory*	*Experiment*
Anthracene	Perylene	2.3	12	3.1	5.4
Perylene	Rubrene	2.8	13	3.8	6.5
9,10-Dichloroanthracene	Perylene	1.7	8.0	4.0	6.7
Anthracene	Rubrene	0.77	3.7	2.3	3.9
9,10-Dichloroanthracene	Rubrene	0.85	3.1	3.2	4.9

Although electric quadrupole interactions with light are several orders of magnitude weaker than electric dipole interactions, the probability of dq energy-transfer (i.e. transfer in which the transition for one partner is dipole-allowed and for the other quadrupole-allowed) is only less by a factor of about 10 than for dd interaction. The theory for the dq process predicts that the strength of interaction should fall off as $1/r^8$, so it is less likely than dd exchange to lead to long-range transfer.

According to the spin selection rule, $\Delta S = 0$, long-range coulombic transfer should be impossible for any process involving multiplicity changes, and long-range triplet–triplet energy transfer would then be excluded. However, to the extent that spin–orbit coupling allows electric dipole optical transitions with $\Delta S \neq 0$ in complex molecules, coulombic transfer can occur by the dd mechanism. Transfer is likely to be slower than for exchange processes in which transitions for donor and acceptor are fully allowed, but, since the actual radiative lifetimes of the triplet states are also long, the long-range energy transfer process may still be important relative to radiation. It follows that the long-range interaction is likely to be demonstrated only in systems in which quenching or ISC is not the predominant loss process for the donor triplet. A very interesting possibility arises that a process such as

$$D^*(T_1) + A(S_0) \rightarrow D(S_0) + A^*(S_1) \tag{5.26}$$

can take place *more* readily than triplet–triplet transfer. The optical transition is only weak for *one* of the exchanging pair, instead of for both, and if radiationless decay of $D^*(T_1)$ is slow, reaction (5.26) can occur via coulombic interaction. In contrast, reaction (5.26) is not favoured by the exchange interaction mechanism since it is non-adiabatic, while triplet–triplet exchange is adiabatic. Triplet–singlet exchange has, in fact, been detected by the emission of *sensitized delayed fluorescence* in rigid glasses of triphenylamine (donor) and chrysoidine (acceptor). The experimental value for r_0 is 5.5 nm, and the theoretical value 4.0 nm. Similar fluorescence sensitized by the triplet of triphenylamine has been detected with chlorophyll and pheophytin as acceptors, and such energy transfer may well be implicated in photobiological processes.

5.4 Excimers and exciplexes

The theme of this chapter has been how intermolecular processes involving excited species can affect the emission of light as a result of the exchange of energy. Another important type of interaction is the formation of relatively long-lived molecular complexes of excited species. Studies of emission provide much information about complex formation, and it is appropriate to introduce the concepts here, although it should be recognized that many other aspects of photochemistry may be influenced as well.

Many molecules that do not interact significantly in their ground states appear to form reasonably stable complexes when excited. The complexes are called *excimers* or *exciplexes*, words derived from **exci**ted **dimer** and **exci**ted com**plex**. Thus, an excimer is produced by the interaction of an excited molecule with a ground-state molecule of the same chemical identity, while an exciplex involves interaction with a chemically different species. Excimers and exciplexes have a fixed and simple stoicheiometric composition, usually 1 : 1.

Excimers were first recognized by the effect of increasing concentration on the fluorescence of some solutes. The intensity of the normal fluorescence decreases, while a new band at longer wavelengths appears, the intensity of which increases with increasing concentration. Some aromatic hydrocarbons such as pyrene show this behaviour particularly clearly. Pyrene itself has a violet fluorescence when it is in dilute solution. At high concentrations, that fluorescence becomes replaced by a structureless blue emission. An excimer is formed between an excited singlet pyrene and ground-state pyrene, and it is the excimer that radiates. If pyrene is represented as P, the excitation scheme is thus

$$P(S_0) + h\nu \rightarrow P(S_1)$$ absorption   (5.27)

$$P(S_1) \rightarrow P(S_0) + h\nu$$ normal   (5.28)
                                                       fluorescence

$$P(S_1) + P(S_0) \rightarrow P(S_1):P(S_0)$$ excimer   (5.29)
                                                       formation

$$P(S_1):P(S_0) \rightarrow P(S_0) + P(S_0) + h\nu$$ excimer   (5.30)
                                                       fluorescence

The emission from the excimer is initially to a $P(S_0):P(S_0)$ pair possessing the same geometry as the excimer, but since there is no attraction between the ground-state molecules it has the same energy as isolated $P(S_0)$ molecules, and immediately separates. The excimer, on the other hand, *is* stabilized with respect to an isolated $P(S_1)$ molecule, and so the emission from the excimer lies at longer wavelengths than the normal fluorescence. Excimer emission is structureless, because the lower state of the transition is essentially a continuum resulting from the repulsion of the ground-state molecules.

Exciplex emission is seen in solutions of mixed solutes. For example, fluorescence of anthracene is quenched by diethylaniline, and a new and structureless emission is observed at longer wavelengths. The emission is not sensitized fluorescence of diethylaniline, but rather emission from the excited singlet anthracene–diethylaniline complex.

Time-resolved studies in both excimer and exciplex systems show clearly how the emitting complex evolves. Following excitation with a flash of light, the emission observed at the shortest times (around 1 ns) is that of the isolated molecule. However, over a period of tens of nanoseconds, the excimer or exciplex emission begins to appear, until, at times of more than about 100 ns after the flash, the emission spectrum is indistinguishable from that seen with steady illumination. The rate at which emission from the complex builds up is close to the rate at which the excited species can diffuse through the solvent towards its partner.

Excimer formation occurs with non-aromatic molecules as well as with aromatic ones, and triplet excimers also exist, giving rise to excimer phosphorescence. The excimer of pyrene is, however, particularly strongly bonded, thermodynamic measurements indicating a bond strength of around $40 \, \text{kJ} \, \text{mol}^{-1}$. An entropy decrease of $80 \, \text{J} \, \text{K}^{-1} \, \text{mol}^{-1}$ on formation of the excimer suggests that the dimer is quite rigid, and it is thought that the aromatic hydrocarbons adopt a sandwich structure with a separation between the planes of about $0.33 \, \text{nm}$.

The nature of the bonding in excimers and exciplexes is clearly dependent on the presence of electronic excitation. Part of the stabilization arises from the promotion of an electron from a filled orbital that would be antibonding in a ground-state dimer or complex to an unfilled bonding orbital in the excited pair. Charge-transfer stabilization is also important in excimers, and especially in exciplexes. Excited species are usually both better electron donors and better electron acceptors than the ground states (cf. Chapter 6). The promotion of an electron to a higher energy level obviously makes an excited molecule potentially a better donor than the ground state, but the lower-energy orbital from which the electron was promoted can also now receive another electron, thus leading to increased acceptor properties. Electrostatic attraction between M^+ and N^- can thus arise in an excited pair $(MN)^*$, the direction of electron transfer depending on the particular chemical species involved. Exciplexes have large dipole moments as a result of the charge-transfer bonding, and the emission spectrum is dependent on the dielectric constant (relative permittivity) of the solvent. Excimers have no dipole moment resulting from their interaction, because the two constituents are the same. However, resonance between M^+M^- and M^-M^+ structures still allows charge transfer to contribute to stabilization.

5.5 Energy-pooling, delayed fluorescence, and 'uphill' processes

Energy may sometimes be transferred from an excited species to an acceptor that is already excited, thus raising the acceptor to a higher electronic state; the process may be referred to as energy-pooling.

We have referred already (Section 4.7, p. 93) to 'dimol' emission from two excited oxygen molecules in the $^1\Delta_g$ state

$$O_2(^1\Delta_g) + O_2(^1\Delta_g) \rightarrow 2O_2 + h\nu_{\lambda = 634 \, \text{nm}} \qquad (5.31)$$

The emitting species appears to be the complex $[O_2(^1\Delta_g) : O_2(^1\Delta_g)]$, and the process is radiative energy-pooling. Further dimol emissions have been observed in the gas-phase oxygen system from the excimers $[O_2(^1\Delta_g) : O_2(^1\Sigma_g^+)]$ and $[O_2(^1\Sigma_g^+) : O_2(^1\Sigma_g^+)]$. Since the emission intensity will be proportional to the product of excited state concentrations, and thus to the square of the absorbed intensity, the quantum yield for the 'dimol' emission intensities will be dependent on the absorbed light intensity.

Molecular, rather than radiative, energy-pooling has been established in many systems. Triplet–triplet pooling to give an excited singlet state is most common, partly because the relatively long lifetime of excited triplets favours the rare triplet–triplet bimolecular process. The reaction

$$D^*(triplet) + A^*(triplet) \rightarrow D(singlet) + A^{**}(singlet) \qquad (5.32)$$

is adiabatic with respect to spin, and for many organic molecules the first excited singlet is energetically accessible from two triplets. An interesting example of a non-adiabatic energy-pooling process is the formation of the second singlet of $O_2(O_2(^1\Sigma_g^+))$ in bimolecular collision of $O_2(^1\Delta_g)$ molecules:

$$O_2^*(^1\Delta_g) + O_2^*(^1\Delta_g) \rightarrow O_2^{**}(^1\Sigma_g^+) + O_2(^3\Sigma_g^-) \qquad (5.33)$$

Emission at $\lambda = 762$ nm (of the forbidden $O_2(^1\Sigma_g^+) \rightarrow O_2(^3\Sigma_g^-)$ system) is observed even though reaction (5.33) is spin-forbidden; the forbiddenness is reflected in the low rate constant for the process ($\sim 10^3$ dm^3 mol^{-1} s^{-1}).

Both donor and acceptor are usually molecules of the same chemical entity, so that reaction (5.32) provides a means of reaching the singlet state when only triplets are present in the system. Energy-pooling between two triplets is known as 'triplet–triplet quenching' or 'triplet–triplet annihilation', and is another mechanism for the emission of *delayed fluorescence* (see also Section 4.6). For example, in anthracene the decay of fluorescence has two components, one with the normal fluorescence lifetime and the other slow, although the spectral distribution of both components is identical. The excitation mechanism (omitting radiationless decay or quenching steps) seems to be

$$A(S_0) + h\nu \rightarrow A^*(S_1) \qquad (5.34)$$

$$A^*(S_1) \rightarrow A(S_0) + h\nu \qquad \text{normal fluorescence} \qquad (5.35)$$

$$A^*(S_1) \rightsquigarrow A^*(T_1) \qquad \text{ISC} \qquad (5.36)$$

$$A^*(T_1) \rightarrow A(S_0) + h\nu \qquad \text{normal phosphorescence} \qquad (5.37)$$

$$A^*(T_1) + A^*(T_1) \rightarrow A^*(S_1) + A(S_0) \qquad \text{energy-pooling} \qquad (5.38)$$

$$A^*(S_1) \rightarrow A(S_0) + h\nu \qquad \text{delayed fluorescence} \qquad (5.39)$$

Reactions (5.35) and (5.39) are, of course, identical, and are written twice to show the sequence of events leading to prompt and delayed fluorescence.

The kind of delayed fluorescence just described does not show the same dependence on temperature as the thermally activated E-type delayed fluorescence (Section 4.6), and it may be distinguished from it by this means. A better distinguishing feature is the emission intensity dependence on the intensity absorbed, which is first-order in E-type delayed fluorescence, but squared in the triplet-annihilation process. Further, E-type delayed fluorescence has the same decay lifetime as that of the triplet–singlet phosphorescence in the same solution; delayed fluorescence excited by the triplet-annihilation mechanism should have a lifetime about one-half of that of the phosphorescence, because of the second-order dependence on triplet concentration.

Triplet-annihilation delayed fluorescence is sometimes known as *P-type delayed fluorescence* because it is observed in solutions of pyrene. However, delayed fluorescence in pyrene shows an additional feature in that the delayed emission appears to derive mainly from the excimer $PP*(S_0S_1)$ [P = pyrene], while the normal, prompt, fluorescence at moderate concentrations shows both monomer and excimer bands. The explanation appears to lie in the mechanism for the triplet–triplet energy-pooling step. If the singlet excimer is a reaction intermediate (related to the transition state), then radiation may occur before the equilibrium concentrations of excimer and monomer can be set up:

$$P*(T_1) + P*(T_1) \rightarrow PP*(S_0S_1) \rightarrow P_2(S_0S_0) + hv \qquad (5.40)$$

and the *delayed* emission will show no component from monomeric excited pyrene, $P*(S_1)$.

Sensitized delayed fluorescence is another process that can be consequent upon energy-pooling. For example, a solution of 10^{-3} mol dm^{-3} of phenanthrene containing 10^{-7} mol dm^{-3} of anthracene shows quite intense delayed emission from the anthracene. This concentration of anthracene is far too low to show direct P-type delayed fluorescence, and the excitation scheme appears to be

$$P(S_0) + hv \rightarrow P*(S_1) \qquad (5.41)$$

$$P*(S_1) \rightsquigarrow P*(T_1) \qquad \text{ISC} \qquad (5.42)$$

$$P*(T_1) + A(S_0) \rightarrow P(S_0) + A*(T_1) \qquad \text{energy transfer} \qquad (5.43)$$

$$A*(T_1) + A*(T_1) \rightarrow A*(S_1) + A(S_0) \qquad \text{energy-pooling} \qquad (5.44)$$

$$A*(T_1) + P*(T_1) \rightarrow A*(S_1) + P(S_0) \qquad \text{energy-pooling} \qquad (5.45)$$

$$A*(S_1) \rightarrow A + hv \qquad \text{delayed fluorescence} \qquad (5.46)$$

In this system it is not possible to distinguish energy-pooling between like (reaction 5.44) and unlike (reaction 5.45) species. However, if a compound whose triplet energy is less than one-half of its excited singlet energy is chosen as acceptor, then triplet-annihilation $A*(T_1) + A*(T_1)$ cannot populate $A*(S_1)$. If the donor triplet level is sufficiently high $D*(T_1) + A*(T_1)$ *can* produce $A*(S_1)$, and it should be possible to observe sensitized delayed fluorescence produced *only* by mixed triplet-annihilation. An example of a suitable donor–acceptor pair is anthracene–naphthacene. Dilute solutions show quite intense delayed emission from naphthacene, although solutions of naphthacene alone show no delayed emission even at higher concentrations.

The energy-pooling process leads to more excitation energy in one product than was present in either reactant, and it is occasionally possible for the radiation ultimately emitted to be of *shorter* wavelength than the exciting radiation. Sensitized 'anti-Stokes' delayed fluorescence has been observed with donor–acceptor pairs such as phenanthrene–naphthalene and pro-flavine–anthracene. The requirement is that the lowest excited singlet of the donor should lie below that of the acceptor, but that the triplet should lie above that of the acceptor (see Fig. 5.3). The observed anti-Stokes emission from naphthalene has 21 kJ mol^{-1}, and that from anthracene 38 kJ mol^{-1}, more energy per quantum than the exciting radiation.

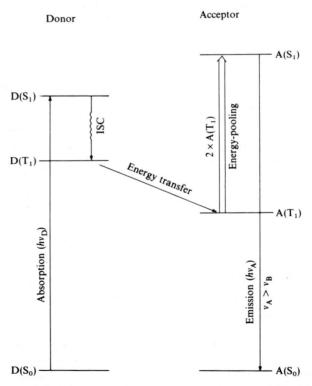

Fig. 5.3 The mechanism of excitation of delayed fluorescence in which the emission is of shorter wavelength than the exciting radiation. Note that D (S$_1$) lies below A (S$_1$), but D(T$_1$) lies above A(T$_1$).

The concentration of the energy from two separately absorbed quanta into a single molecular species is demonstrated most dramatically in the sensitized anti-Stokes emission experiments, although it occurs, in principle, in all energy-pooling processes. The explanation of this apparent contravention of the Planck energy–frequency relation (and of the Stark–Einstein law) may provide an understanding of the primary processes in photosynthesis, in which just such a concentration of photon energy is needed for the occurrence of photochemical reactions.

Photon energy may occasionally be upgraded by true *biphotonic absorption* (as distinct from tandem excitation of one level and then another). Multiphoton absorption processes offer an interesting mechanism for the occurrence of reactions that are at first sight photochemically impossible (although it is unlikely that they are important in any *terrestrial* natural phenomena). As we explained in Section 3.9, the high intrinsic intensity of laser radiation makes possible the 'simultaneous' absorption of two photons, and emission phenomena involving double quantum excitation have been observed. For example, emission in caesium vapour from the $9^2D_{3/2} \rightarrow 6^2P_{3/2}$ transition ($\lambda = 584.7$ nm) may be excited by radiation from a laser tuned to $\lambda = 693.78$ nm, although the caesium is transparent to red light of that wavelength under normal conditions. However, $\lambda = 693.78$ nm corresponds to exactly half the energy required to excite the $9^2D_{3/2}$ state of caesium from the ground, $6^2S_{1/2}$, state ($28\,828$ cm^{-1}), and the upper state appears to be populated by two-photon absorption. Similar observations have been made of fluorescence of aromatic hydrocarbons (phenanthrene, anthracene, pyrene, benzpyrene) excited by laser radiation of a wavelength supposedly not absorbed by the molecules. It can be shown that the effect does not result from frequency doubling of the incident radiation, nor from successive excitation.

5.6 Photosensitized reactions

Emission is not the only possible fate of the electronically excited acceptor formed in intermolecular energy transfer. Dissociation and chemical reaction are two other important routes for utilization of electronic excitation. These processes are discussed in more detail in Chapters 3 and 6, but it is of interest to examine here the occurrence of *photosensitized reactions* in terms of the principles of energy transfer. Such reactions take place in a species other than the one absorbing the radiation, and sensitization often makes it possible to induce photochemical change in a wavelength region where the reactant is transparent. For example, CH_4 is virtually transparent at $\lambda > 170$ nm. However, in the presence of mercury vapour, the mercury resonance line at

$\lambda = 253.7$ nm will dissociate CH_4. The sequence of events can be written

$$Hg + h\nu_{\lambda = 253.7 \text{ nm}} \rightarrow Hg^* \tag{5.47}$$

$$Hg^* + CH_4 \rightarrow CH_3 + H + Hg \tag{5.48}$$

with energy of the excited mercury atom leading to the dissociation of methane.

Energy transfer may permit the population of electronic *states* of molecules different from those populated by absorption, and photosensitized processes may differ chemically from unsensitized photolyses. The direct photolysis ($\lambda < 144$ nm) of methane, for example, yields molecular hydrogen and methylene (CH_2) as the primary products, while the mercury-sensitized photolysis ($\lambda = 253.7$ nm) gives mainly CH_3 and H atoms, as indicated in eqn (5.48).

The quantum yield for the direct photolysis at $\lambda = 313$ nm of ethyl pyruvate (ethyl 2-oxopropanoate)

$$CH_3COCOOC_2H_5 + h\nu \rightarrow 2CH_3CHO + CO \tag{5.49}$$

is 0.17 in benzene solution. However, in the presence of benzophenone, but at the same wavelength, the quantum yield for ethyl pyruvate removal is 0.32. It is thought that the triplet state of the pyruvate, not rapidly populated from the singlet, is produced by energy transfer from triplet benzophenone, and that it is from this state that dissociation occurs. The transfer is energetically favourable (triplet energies for benzophenone and ethyl pyruvate are 289 kJ mol^{-1} and 272 kJ mol^{-1}, respectively); it is interesting that 2-aceto-naphthone, with a triplet energy (247 kJ mol^{-1}) *below* that of the pyruvate, is ineffective as a sensitizer. The energy transfer process is assumed to be adiabatic and, hence, spin-conserved, in systems such as the one described, which employ organic sensitizers, and the occurrence of photosensitized reactions provides valuable information about intersystem crossing and triplet chemistry. We shall discuss this application further at the end of the section.

The question of spin conservation in collisional energy transfer, and, hence, in photosensitized reaction, is a vexed one, but of importance because of the possibility that specific spin multiplicity could be expected in products of sensitized reactions. The problem is that the Russell–Saunders spin, S, cannot be a perfect quantum number, even in the isolated reaction partners, because radiative and radiationless triplet–singlet transitions occur, yet it must be supposed that S is a good enough quantum number to make spin-forbidden collisional *reaction* an inefficient process. However, the assumption of spin conservation is often made, and since experimental results are not often inconsistent with it, we shall imply that spin is conserved in the rest of the discussion. More certainty attaches to experiments where only a triplet state can be populated because of the energies involved (cf. Fig. 5.2 and the associated discussion of sensitized phosphorescence).

Many substances have proved popular as sensitizers. For triplet sensitization studies, ketones have several desirable features. High triplet yields are obtained in carbonyl compounds, while the small singlet–triplet splitting, and relatively high triplet energies, means that the energy relations implied by Fig. 5.2 can often be achieved. Benzophenone is frequently used in condensed-phase experiments, while biacetyl is particularly useful for gas-phase work. Historically, mercury was much employed in the gas phase because of its volatility at room temperature, and the ease with which emission lamps for the $\lambda = 253.7$ nm resonance line can be constructed. Other volatile metals have been used as photosensitizers, among them cadmium, zinc, thallium, indium, calcium, sodium, and gallium; the rare gases are useful sensitizers in the vacuum ultraviolet region.

One method for the study of triplet states, which depends on photosensitization, is due to Hammond and co-workers. The technique consists in part of using the rate of an isomerization of an excited triplet as a measure of the rate at which the triplet is populated (e.g. the rate of *cis–trans* isomerization in penta-1,3-diene, 1,2-dichloroethene, or 2-pentene may be used as an indicator). The triplet of an acceptor may be populated by energy transfer from a donor triplet, and if the singlet of the acceptor lies *above* the donor levels (as in Fig. 5.2), the only excited state of the acceptor that can be populated is its triplet. If $D^*(T_1)$ lies at a sufficiently high energy above $A^*(T_1)$, energy transfer is diffusion-controlled (see Section 5.2), and at moderate A concentrations the rate of the isomerization is the same as the rate at which $D^*(T_1)$ is populated, so long as intermolecular 'forbidden' singlet–triplet transfer from $D^*(S_1)$ to $A^*(T_1)$ is negligibly slow in comparison. Hence, the rate, or quantum yield, of intersystem crossing in the *donor* $D^*(S_1) \rightsquigarrow D^*(T_1)$ may be calculated.

The method may also be used to obtain an idea of the relative energies in triplet levels of donor and acceptor, and thus fix an unknown triplet energy if the other is known from spectroscopic data. The isomerization reaction becomes slower as the transfer decreases in efficiency when ΔE for $D^*(T_1) - A^*(T_1)$ becomes positive. The fall in rate may not bear an exponential relationship to positive ΔE (see Section 5.2, p. 108), but it is still possible to obtain a measure of the relative triplet energies.

5.7 Lasers, stimulated emission, and population inversions

Lasers are now of widespread use in the sciences and industry, and they are beginning to make a significant impact in domestic life with typical applications in scanners at supermarket checkouts, and in video and compact-disc reproduction. The study of photochemistry has itself benefited much from the special properties of laser radiation, such as monochromaticity, high intensity, and short-pulse duration. Laser methods contribute to the techniques available to the experimental photochemist, and they are accordingly discussed in Chapter 7 of this book. That laser action is possible at all is, in many types of laser, a result of the operation of processes of the kind that we have been discussing in this chapter and the preceding one. It is appropriate, therefore, to conclude this chapter with a brief survey of some important classes of laser in terms of the underlying photochemistry.

Lasers depend, as their name (**L**ight **A**mplification by **S**timulated **E**mission of **R**adiation) implies, on the production of stimulated emission from an excited system, rather than the spontaneous emission that we have been considering so far. *Net* stimulated emission is observed only in systems where the population of the excited state is greater than that of the ground state, a situation described as a population inversion (cf. Section 2.3, p. 18). Our main concern here is to investigate ways in which the inversion may be achieved, but we must first understand the basic principles of laser action.

Figure 5.4 shows the essential features of a laser system. The laser medium, which contains the excited species with the inverted population, is in the form of a cylinder or contained within a cylindrical tube. The medium is enclosed in an optical cavity, usually made up of a totally reflecting mirror at one end, while the other endplate is made partially transmitting to allow some light to escape. Laser action can be imagined to start with the production of a photon by the spontaneous emission mechanism. As that radiation traverses the medium, it stimulates emission of further photons, and the intensity of the radiation is increased, i.e. it is *amplified*. The optical cavity is made to be resonant at the wavelength of the radiation; the intensity therefore builds up as the radiation is reflected back and forth. Emission will cease when the radiative loss has been sufficient to destroy the initial population inversion, and a *pulse* (or flash) of radiation is obtained. *Continuous-wave* (*CW*) lasers are provided with some way of replacing the excited species as fast as they are lost by stimulated emission of radiation.

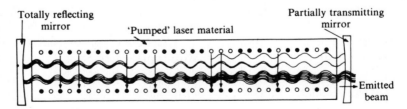

Fig. 5.4 Representation of a laser system. Open circles represent the upper state of the emitting species, and solid circles the lower state. Induced radiation builds up in intensity as the wave travels to and fro in the cavity between the mirrors. (Based in part on Pimentel, G. C. (1966). *Sci. Am.* **214**(4), 32.)

We turn now to a consideration of methods by which population inversions are attained in some lasers of practical importance. Intra- and intermolecular energy transfer processes play an important part in these mechanisms. Thermal excitation cannot, by definition, lead to inversion in an equilibrated system. Nor can direct absorption of light give an inversion in a simple system consisting of two levels, because the incident ('pumping') radiation will not only excite the lower state to the upper, but also promote stimulated radiation from the upper to depopulate it. A *three-level* system can, however, be used to obtain a population inversion and, hence, laser action. Figure 5.5(a) shows the operation of a three-level system, and is a simplified form of the system applicable to the *ruby laser* (ruby is a crystalline aluminium oxide with some of the aluminium ions replaced by chromium). The optical pumping is usually supplied in the form of a flash, and excites the upper level of the absorption band, C. Radiationless transition populates the level B during the course of the pumping flash, and a population inversion of B with respect to X can be built up. The essential element in the process is that the C→B radiationless transition should be *rapid*, and the B→X radiative transition relatively slow, so that the population of B can be increased at the expense of C. More than one-half of X must be excited to B (via C) before laser action can occur. In spite of this constraint on three-level systems, continuously-operating ruby lasers can be constructed, although it is difficult to obtain the very high pumping light intensities needed, and it may be necessary to cool the crystal to remove heat liberated in the radiationless energy transfer.

A four-level system, as in Fig. 5.5(b), is potentially much more efficient than the three-level system. Various lasers based on neodymium, such as Nd-doped glass or Nd-doped yttrium aluminium garnet (*Nd–YAG laser*) are of

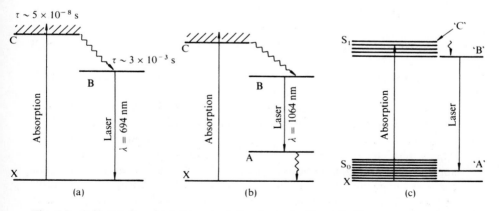

Fig. 5.5 Schemes for obtaining population inversion in optically pumped systems: (a) three-level (ruby laser); (b) four-level (Nd laser); (c) four-level (dye-laser).

this kind. High pulse and CW powers are available from these lasers. The new feature in Fig. 5.5(b) compared with Fig. 5.5(a) is the level A to which the emission from B occurs. Since this level is initially unpopulated, there is no need for large fractions of X to be excited to C for [B] to be larger than [A] and for laser action to start. For CW operation, it is necessary, of course, that the A state is rapidly depopulated (by radiationless transitions in solid lasers) in order to maintain the population inversion with respect to B.

Organic *dye lasers* are of the four-level type just described. Many dyes (e.g. based on rhodamine or coumarin) can be made to produce laser radiation on optical excitation, which is often achieved with another 'pump' laser. Although only two electronic levels may be involved directly, the existence of the vibrational sublevels within the electronic states makes these lasers formally four-level systems, as indicated in Fig. 5.5(c). Collisional quenching of A can maintain the population inversion with respect to B in liquid-phase dye lasers. The broad-band absorption and emission characteristics of the dyes themselves allow *tuneable* lasers (cf. Section 7.2) to be built, and dyes are commercially available that encompass the spectral range 200–1100 nm.

Gas-phase lasers usually employ electrical excitation to obtain population inversions. Collisions with energetic electrons can excite (and ionize) molecules and atoms, and the optical selection rules do not apply so that 'forbidden' metastable states become accessible. Transitions to lower-lying states may then make laser action possible. Two important examples are the *nitrogen laser*, which is a pulse laser, and the *argon ion laser*, which is generally operated as a CW device. The relevant processes are described by reactions (5.50)+(5.51) and (5.52)+(5.53)

$$N_2(X^1\Sigma_g) \overset{e}{\to} N_2(B^3\Pi_u) \tag{5.50}$$

$$N_2(B^3\Pi_u) \to N_2(C^3\Pi_g) + h\nu(\lambda = 337.1 \text{ nm}) \tag{5.51}$$

$$Ar(3s^2 3p^6) \overset{e}{\to} Ar^+(3s^2 3p^5) \overset{e}{\to} Ar^+(3s^2 3p^4 4p) \tag{5.52}$$

$$Ar^+(3s^2 3p^4 4p) \to Ar^+(3s^2 3p^4 4s) + h\nu(\lambda = 488.0, 496.5, 514.5 \text{ nm}) \tag{5.53}$$

Some very important gas lasers rely on intermolecular energy transfer processes in the excitation mechanism. For example, in the *helium–neon laser* an electric discharge is passed through a mixture of about 10% Ne in He. Impact with electrons from the discharge excites the first excited triplet and singlet states of He, as indicated in Fig. 5.6. Optical transitions from these states to the ground states are forbidden, and they are therefore metastable and long-lived. The two states are nearly resonant with two excited states of Ne (designated 2S and 3S in the figure), and efficient collisional energy exchange leads to production of excited neon in the S states. There are

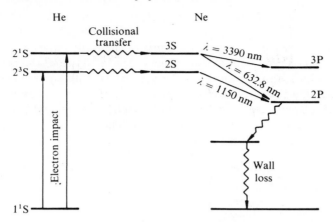

Fig. 5.6 Energy levels and transitions in the helium–neon laser.

lower-lying P states for which efficient resonant excitation is not possible, so that the conditions exist for a population inversion of the S states with respect to the P states. One of the laser transitions (designated $3s_2 \rightarrow 2p_4$) is the familiar red line at $\lambda = 632.8$ nm.

The *carbon dioxide laser* uses near-resonant *vibrational* energy transfer to obtain a population inversion of the emitting levels. Here, a CO_2–N_2–He mixture is used, and an electric discharge produces $N_2(v=1)$. The excited nitrogen is almost resonant with the v_3 vibration of CO_2, and efficient energy transfer to CO_2 takes place (note the nomenclature $CO_2(v_1, v_2, v_3)$, where v_1, v_2, v_3 are the vibrational quantum numbers of the symmetric stretch, bending, and asymmetric stretch vibrations, respectively).

$$N_2(v=1) + CO_2(000) \rightarrow CO_2(001) + N_2(v=0) \qquad (5.54)$$

$$CO_2(001) \rightarrow CO_2(100) + hv(\lambda = 10.6 \ \mu m) \qquad (5.55)$$

$$CO_2(001) \rightarrow CO_2(020) + hv(\lambda = 9.6 \ \mu m) \qquad (5.56)$$

The population of $CO_2(001)$ is inverted with respect to both (100) and (020) levels, and laser action can occur. The helium in the mixture serves both to increase the efficiency of vibrational excitation of N_2 and to deplete the lower levels produced by the CO_2 emission, thus giving high powers and allowing CW operation.

The discussion of excimers and exciplexes in Section 5.4 indicates yet another way in which a population inversion may be obtained. Since the ground state of a pair that form an excited-state complex is normally repulsive, the lifetime of the ground state is around one vibrational period and the population is negligible. Formation of the excited-state complex necessarily gives it a higher population than the hypothetical ground state, and laser operation is possible. *Excimer lasers* work in this way, although the name should really be 'exciplex' for some of the most important examples that are based on rare-gas–halogen systems. Argon, krypton, and xenon form exciplexes with F and Cl atoms (and Xe with Br as well). Laser emission may be obtained, the shortest wavelength being for ArCl at $\lambda = 175$ nm, well into the vacuum ultraviolet region. The initial excitation is once again in the form of an electric discharge, and the reaction sequence is

$$\mathrm{Ar} \xrightarrow{\mathrm{e}} \mathrm{Ar}^* \tag{5.57}$$

$$\mathrm{Ar}^* + \mathrm{F}_2 \rightarrow \mathrm{ArF}^* + \mathrm{F} \tag{5.58}$$

$$\mathrm{ArF}^* \rightarrow \mathrm{Ar} + \mathrm{F} + h\nu(\lambda = 193 \text{ nm}) \tag{5.59}$$

Chemiluminescent systems can exhibit stimulated emission if a population inversion of the excited states can be produced. Several *chemical lasers* based on excitation by chemical reaction have been demonstrated. For example, vibrationally excited HCl is generated by the reaction between atomic hydrogen and Cl_2, and lasers have been made that employ this system. One reason why much interest attaches to chemical lasers is the extremely large energy releases (of the order of kilojoules per mole) that are available in rapid chemical reactions.

Bibliography

Parker, C. A. (1969). *Luminescence of solutions.* Elsevier, Amsterdam.

Turro, N. J. (1978). *Modern molecular photochemistry*, Chapter 9. Benjamin/Cummings, Menlo Park, CA.

Calvert, J. G. and Pitts, J. N., Jr (1966). *Photochemistry*, Chapter 4. John Wiley, Chichester and New York.

Barltrop, J. A. and Coyle, J. D. (1978). *Principles of photochemistry*, Chapter 4. John Wiley, Chichester and New York.

Bowen, E. J. (ed.) (1968). *Luminescence in chemistry.* Van Nostrand, Princeton, NJ.

Wilkinson, F. (1964). Electronic energy transfer between organic molecules in solution. *Adv. Photochem.* **3**, 241.

Krause, L. (1975). Sensitized fluorescence and quenching. *Adv. Chem. Phys.* **28**, 267.

Special topics
Section 5.4

Kuzmin, M. G. and Soboleva, I. V. (1986). Exciplexes in photoreaction kinetics. *Prog. React. Kinet.* **14**, 157.

Gordon, M. and Ware, W. R. (eds) (1975). *The exciplex.* Academic Press, New York.

Stevens, B. (1971). Photoassociation in aromatic systems. *Adv. Photochem.* **8**, 161.

Klopffer, W. (1973). Intramolecular excimers. In *Organic molecular photophysics* (ed. J. B. Birks) Vol. 1. John Wiley, Chichester and New York.

Beens, H. and Weller, A. (1975). Excited molecular pi-complexes in solution. In *Organic molecular photophysics* (ed. J. B. Birks) Vol. 2. John Wiley, Chichester and New York.

Lim, E. C. (1987). Molecular triplet excimers. *Acc. Chem. Res.* **20**, 8.

Yakhot, V., Cohen, M. D., and Ludmer, Z. (1979). What's new in excimers. *Adv. Photochem.* **11**, 489.

Section 5.6

Gunning, H. E. and Strausz, O. P. (1963). Isotope effects and the mechanism of energy transfer in mercury photosensitization. *Adv. Photochem.* **1**, 209.

Kavarnos, G. J. and Turro, N. J. (1986). Photosensitization by reversible electron transfer: theories, experimental evidence, and examples. *Chem. Rev.* **86**, 401.

Engel, P. S. and Monroe, B. M. (1971). Complications in photosensitized reactions, *Adv. Photochem.* **8**, 245.

Section 5.7

Svelto, O. (1982). *Principles of lasers* (translated by D. C. Hanna). Plenum Press, New York.

Siegman, A. E. (1986). *Lasers.* Oxford University Press, Oxford.

Fleming, G. (1982) Lasers. In *Light, chemical change and life* (eds J. D. Coyle, R. R. Hill, and D. R. Roberts). Open University Press, Milton Keynes.

Phillips, R. (1983). *Sources and applications of ultraviolet radiation*, Chapter 12. Academic Press, New York.

Rabek, J. F. (1982). *Experimental methods in photochemistry and photophysics*, Chapter 16 (Part 2). John Wiley, Chichester and New York.

Pimentel, G. C. (1966). Chemical lasers. *Sci. Am.* **214**(4), 32.

Lin, M. C., Umstead, M. E., and Djeu, N. (1983). Chemical lasers. *Ann. Rev. Phys. Chem.* **34**, 557.

Burdett, J. K. and Poliakoff, M. (1974). Tunable lasers. *Chem. Soc. Rev.* **3**, 293.

Snavely, B. J. (1973). Organic dye lasers. In *Organic molecular photophysics* (ed. J. B. Birks) Vol. 1. John Wiley, Chichester and New York.

6

Reactions of excited species

6.1 Introduction

The chemistry of an excited species may differ markedly from that of the ground-state species, and, as we pointed out in Chapter 1, the differences may come about both as a result of the excess energy carried by the excited species and as a result of the particular electronic arrangement of the excited state. Both factors appear clearly in the intramolecular and intermolecular transfer of energy discussed in the last two chapters. Excess energy is obviously a prerequisite for its transfer, and the restriction on the electronic states between which energy may be transferred is a consequence of the manner in which the electrons are arranged in the various states. In the present chapter we shall consider processes, involving excited species, which lead to *chemical* reaction (i.e. in which the reactants and products differ in chemical identity rather than state of excitation). These chemical processes can be either intramolecular or intermolecular in the same way as the physical process of energy transfer. The first class of reactions includes intramolecular reductions, additions, and various types of isomerization; intermolecular reactions of excited species include those with added reactants, with unexcited molecules of the absorbing substance, or, in solution, with the solvent. Photochemical reactions may offer the best synthetic routes to a variety of curious, interesting, or useful compounds: some examples are given in Section 8.10. We describe here some of the principles that govern the reactivity of excited species, and present a small selection of reactions that illustrate some of the more important types of process known.

6.2 Reactivity of excited species

In this section we shall consider three factors that may contribute to the apparent reactivity of an excited species: (i) the intrinsic reactivity of the specific electronic arrangement; (ii) the effect of the excitation energy; and (iii) the lifetime of the particular excited state.

Atoms or molecules might be expected to show differing reactivity according to the way in which the electrons are distributed in the available orbitals, and, indeed, differences in reactivity for different states may frequently be shown experimentally. For example, following absorption, most aromatic carbonyl compounds undergo rapid intersystem crossing to the lowest triplet. In 'normal' compounds such as benzophenone, this triplet is (n, π^*) in character, although for some 4-substituted ketones (e.g. 4-aminobenzophenone) the phosphorescence and ESR spectra suggest that the lowest triplet is either (π, π^*) or a charge–transfer state. The reactions of 'normal' and 'abnormal' compounds are entirely different. Triplet benzophenone abstracts hydrogen from suitable solvents, and adds across double bonds. Triplet 4-aminobenzophenone does not participate efficiently in either reaction. It is no great surprise, of course, that promotion of a non-bonding electron, essentially localized on the carbonyl oxygen atom, produces a species whose reactivity is not the same as that resulting from promotion of a π electron spread over the $>C=O$ group.

Electronic excitation may alter the intrinsic reactivity of a species through mechanisms related to the wave (i.e. non-classical) behaviour of the electrons, and this aspect will be explored in greater detail in the next section. There are, however, some general reasons why excited states behave differently from their unexcited parents. The influences include alterations in (i) geometry, (ii) dipole moment, (iii) electron donating and accepting ability, and (iv) the related acid–base properties.

Excitation may alter both the sizes and the shapes of molecules. For a particular reaction, the new steric arrangements may then increase (or, indeed, decrease) reactivity. Movement of electrons between bonding, non-bonding, and anti-bonding orbitals may be expected to change the molecular dimensions. Shapes of molecules may be affected by changes in the nature of the bonding. For example, while the ground state of an alkene such as ethene is planar, the equilibrium structure of the (π, π^*) excited state has the two CH_2 groups lying in perpendicular planes. Promotion of the electron is accompanied by uncoupling of the π orbital to leave only a σ-bond between the carbon atoms. The two electrons in the 2p orbitals on the C atoms experience minimum electrostatic repulsion with the perpendicular geometry. The lengths of the C–C bond are 0.134 nm in the ground state, 0.144 nm in the S_1 state, and 0.158 nm in the T_1 state (also twisted): for comparison, it is worth noting that the C–C bond length in ethane is 0.154 nm.

Dipole moments are influenced by electronic excitation, not only through changes in geometry of the molecular skeleton but also through the redistribution of the electrons themselves. By indicating the distribution, the dipole moments thus suggest the possible chemical behaviour of excited states. Changes in dipole moment on excitation may be recognized by the effects of polar solvents on absorption and fluorescence spectra, and by the effects of applied electric fields on the depolarization of fluorescent emission excited by polarized light. Both increases and decreases are known. In formaldehyde (methanal), for example, the dipole moment decreases from 2.3 Debye in the ground state to 1.6 Debye in the (n, π^*) state, while for benzophenone the values are 2.9 Debye and 1.2 Debye for ground and excited states. The decrease in dipole moments indicates reduced polarization in the $C = O$ bond in the excited molecule. On the other hand, the dipole moment of an aromatic molecule such as 4-nitroaniline increases from 6 to 14 Debye on excitation. There is thus a considerable degree of charge-transfer in the excited state: the fully dipolar structure of 4-nitroaniline, with full negative charges on each oxygen and full positive charges on each nitrogen, would be expected to have a dipole moment of about 25 Debye.

The large increase in dipole moment on excitation of 4-nitroaniline is, in part, a consequence of the molecule possessing an electron-donor and an electron-acceptor group linked through a conjugated system. Excited species are usually *both* better electron donors *and* better electron acceptors than the ground-state species. Excitation in most molecules involves promotion of a (paired) electron from a lower to a higher orbital. The electron that has been promoted is thus more easily removed (ionized) than it was in the ground-state species. At the same time, the promotion of the electron leaves behind a low-lying vacancy that can accept another electron. The excited molecule can both give up its excited electron, and accept an electron into the hole. Put another way, the ionization energy of the molecule has been reduced, and the electron affinity increased, as a result of excitation. The redox properties of excited species are thus likely to be very different from those of unexcited species. Indeed, complete electron transfer to or from an excited molecule is quite common, and many photochemical reactions involve radical anions or cations (cf. particularly the discussion of photosynthesis in Section 8.3).

As a consequence of changed donor–acceptor properties, it might be expected that acid–base behaviour would also be influenced by electronic excitation. Experimental evidence from the pH-dependence of fluorescence spectra shows that the S_1 states of molecules may be either stronger or weaker acids than the ground, S_0, state. For example, for phenols, the pK_a value may be 6 units smaller for S_1 than for S_0 (i.e. S_1 is 10^6 times as acidic), while for aromatic carboxylic acids, the pK_a values can be up to 8 units larger in the excited state (i.e. S_1 is 10^8 times more *basic* than S_0). In the excited phenols, proton exchanges appear to be so rapid that an acid–base equilibrium is set up between excited phenol, HA^*, and H^+ plus the conjugate base, A^-*. The differences in emission maxima in the fluorescence spectra of HA and A^- may be used to estimate the free energy difference between S_1 of HA and S_1 of A^-, and hence the expected pK_a values of the excited state. The results correspond well with direct determinations of pK_a based on intensity measured as a function of pH. The thermodynamic cycle employed in the calculation is known as a *Förster–Weller cycle*. The greatly increased acidity of the excited phenol is attributed to the transfer of electrons from the hydroxy oxygen to the aromatic nucleus, leading to increased positive charge on the oxygen. Triplet states display pK_a values comparable with those of the ground singlet states. It appears that highly charged structures in which two electrons are transferred to the nucleus are excluded, because spin correlation keeps the electrons apart in triplets. This explanation is straying into wave mechanics, but it does emphasize the different reactivity that may be shown by triplet and singlet states: the singlet here behaves as a zwitterion, and the triplet more like a diradical.

Excited singlet and triplet states of organic compounds may react in distinct ways (although to observe the behaviour of the singlet species, it is necessary that chemical reaction proceeds more rapidly than the intersystem crossing from S_1 to T_1). For example, the direct photoisomerization of *trans*-penta-1,3-diene gives different products from the sensitized process

$$\text{Direct} \tag{6.1a}$$

$$\text{Sensitized} \tag{6.1b}$$

The first process involves the rearrangement of an excited singlet reactant, while the second involves the triplet. Another similar example will be encountered in Section 6.4, reactions (6.10a) and 6.10(b).

Another way in which the reactivity of a species may be affected by its multiplicity is illustrated by the triatomic molecule methylene, CH_2. Flash photolysis studies indicate that the ground state is a *triplet*, although there is a first excited singlet lying not far above the ground state (excitation energy ca. $30\,\text{kJ mol}^{-1}$). Singlet methylene is a product (possibly together with some triplet CH_2) of the photolysis of CH_2N_2 or of CH_2CO. Intersystem crossing from the singlet to the triplet is induced by inert gases. The chemical reactivities of the triplet and singlet species are quite different. The singlet state reacts three orders of magnitude more rapidly with H_2 and with CH_4 than does the triplet. Singlet CH_2 inserts into the $C-H$ bonds of alkanes, while the triplet abstracts H atoms

$$^1CH_2 + RH \rightarrow R\text{-}CH_3 \tag{6.2}$$

$$^3CH_2 + RH \rightarrow CH_3 + R \tag{6.3}$$

Methylene adds to substituted ethenes to yield substituted cyclopropanes

$$R_1CH = CHR_2 + CH_2 \rightarrow R_1C \underset{\underset{H_2}{C}}{\diagdown \qquad \diagup} CR_2 \tag{6.4}$$

Singlet CH_2 yields a product that predominantly retains the geometrical configuration of the reactant. *cis*-1,2-Dimethylcyclopropane, for example, is favoured in the reaction of $^1CH_2$ with *cis*-but-2-ene. Stereospecific behaviour of this kind does not seem to be exhibited by $^3CH_2$. One explanation is that the singlet species adds in a single *concerted step* across the double bond, while the triplet first forms a diradical which then ring-closes

$$^3CH_2 + R_1CH{=}CHR_2 \rightarrow \left[\begin{array}{c} R_1CH-\dot{C}HR_2 \\ | \\ \dot{C}H_2 \end{array} \right] \rightarrow \begin{array}{c} R_1C-CR_2 \\ \diagdown\diagup \\ CH_2 \end{array} \qquad (6.5)$$

Rotation about the single C–C bond in the intermediate results in loss of specificity.

We consider now the influence that the energy possessed by an excited species may have on reactivity. Endothermic reactions require that heat be supplied to the system if the process is to occur spontaneously; if isolated from an external heat source, the system will cool down and reaction will become progressively slower. Even if it is possible to supply energy to the reacting species, the activation energy for reaction must be greater than the heat of reaction. It may be expected, therefore, that highly endothermic reactions will proceed only very slowly at room temperature. However, the excess energy carried by an excited reactant may be able either to contribute to the kinetic energy needed to overcome the activation barrier, or to participate in a reaction on a different potential surface for which the barrier height between reactants and products is smaller than the ground-state activation energy.

Even exothermic reactions can possess appreciable activation energies, and energy-rich species might be expected to exhibit enhanced reactivity in such processes. In the few instances where rates of reaction of both ground and excited states have been measured, the energy-rich species is the more reactive. For example, in the decomposition of ozone by atomic oxygen

$$O + O_3 \rightarrow 2O_2 \qquad (6.16)$$

the reaction may be four orders of magnitude faster with excited $O(^1D)$ (process $582\,kJ\,mol^{-1}$ exothermic) than with ground-state $O(^3P)$ (process $389\,kJ\,mol^{-1}$ exothermic; $E_a \sim 23\,kJ\,mol^{-1}$). Since the rate with $O(^1D)$ as reactant approaches the gas-kinetic collision rate, it would appear that the $192\,kJ\,mol^{-1}$ excess energy in $O(^1D)$ can overcome the $123\,kJ\,mol^{-1}$ activation barrier.

In most organic molecules, the ground state is a singlet, and the triplets are necessarily energy-rich. There is, in fact, a general belief that the triplet state of organic molecules is the one that always participates in excited-state reactions because of its high reactivity. This view seems, however, to be ill-founded. Although there may sometimes be a definite dependence of reaction path on multiplicity, as in some of the examples cited earlier in the section, in many instances both singlet and triplet participate in the same reactions. Further, the excited singlet *may* be more reactive than the triplet. Thus it would, perhaps, be more accurate to say that in many photochemical reactions of organic species the triplet is not innately more reactive than the singlet, but it contributes more than the excited singlet to the overall reaction. Even if a triplet is less reactive than a singlet state, it may survive radiative, radiationless and collisional quenching so much better than the singlet that it has a greater probability of undergoing the chemical reaction. In other words, the rate constant for reaction *relative* to that for loss processes may be higher for T_1 than for S_1, even if the absolute rate constant for reaction is lower for the triplet. Kinetic arguments may, in fact, be used to 'prove' the participation of the triplet state. For example, let us consider the photoreduction of benzophenone in the presence of a suitable hydrogen donor, RH,

$$C_6H_5COC_6H_5 + h\nu \xrightarrow{\text{RH}} C_6H_5C(OH)C_6H_5 \qquad (6.7)$$

(This photoreduction reaction will be described further in Section 6.6.) With a hydrogen donor such as $C_6H_5CH(OH)C_6H_5$, at 0.1 mol dm^{-3} concentration, the quantum yield for benzophenone disappearance, ϕ_B, is nearly unity. This at once excludes the excited singlet of benzophenone as the reactive species. The rate constant for H atom abstraction by the singlet must be less than 10^8 dm^3 mol^{-1} s^{-1}, since physical quenching of S_1 proceeds at least 100 times faster, and yet is diffusion-controlled (i.e. $k_q \not> 10^{10}$ dm^3 mol^{-1} s^{-1}). At the same time, the rate constant for $S_1 \rightsquigarrow T_1$ intersystem crossing in benzophenone is about 10^{10} s^{-1}, so that competition between hydrogen abstraction and ISC places a limit on ϕ_B of $(10^8 \times 0.1/10^{10})$ $= 10^{-3}$ at [RH] $= 0.1$ mol dm^{-3} for reaction of S_1. On the other hand, loss processes for T_1 are much slower than those for S_1 (e.g. the rate constant for $T_1 \rightsquigarrow S_0$ ISC in benzophenone is about 10^5 s^{-1}), and reaction competes effectively with the other processes. Further confirmation that the triplet is the important reactive species comes from a comparison of rate data obtained explicitly for the triplet with that obtained from the kinetic dependences of ϕ_B for the 'unknown' state involved in the photoreduction. The benzophenone triplet has been identified in flash photolysis experiments, and the individual rate constants for quenching, ISC, and H atom abstraction are virtually identical with those derived for excited benzophenone from the data for ϕ_B; thus, the excited benzophenone is very probably in the triplet state.

6.3 Correlation rules and symmetry conservation

In this section we shall explore some rationalizations of the reactivity of electronically excited states that seek to explain observed interactions on the basis of the particular electronic arrangement in the excited species. We have seen in Section 5.2 how the collisional transfer of energy might be efficient only if the process can occur adiabatically on a continuous potential surface that links reactants with products. The reactants and products are said to be correlated in this case. Electron spin correlation rules are of particular importance, and, to the extent that the quantum number S is a good description of the system, total electron spin is conserved. Such considerations lie behind the belief that the triplet state of a sensitizer molecule like benzophenone excites a triplet of an acceptor molecule, although the energetics of the system may also demand that the triplet rather than the singlet is produced (cf. p. 118). Identical arguments apply to the conservation of spin in reactions such as addition, abstraction, or exchange, where chemical change occurs. The rule does not say that the reaction *will* occur, but only that it is not prevented from occurring by quantum mechanical constraints. Other factors, such as high activation energies or excessive geometrical distortion, may make the adiabatic reaction improbable. All the further conservation and correlation ideas to be developed in this section have in common this feature of *permitting* but not *ensuring* reaction.

Other quantum numbers may need to be conserved for reaction to be adiabatic, depending on whether they retain any meaning in passage from reactants, through the transition state, to the products. For instance, in the co-linear exchange of an atom with a diatomic molecule, $A + B-C \rightarrow A-B + C$, conservation of orbital electronic angular momentum as well as of spin would be needed to ensure adiabatic correlation of reactants and products. In the more general case, where the molecular symmetry is lower, it may only be possible to specify the conservation of all (or even just some) of the symmetry of the electronic states of the reactants in passing through the transition state to the products. The application of these principles is well established in the reactions of atoms and small molecules, but relies heavily on group-theoretical methods. Some very important classes of organic reaction can also be treated by similar methods, and our concern here is to present a qualitative picture of how conservation of symmetry affects the outcome of these processes.

One of the most fruitful developments in the understanding of organic reaction mechanisms over the last decades has been in the field of *pericyclic* reactions. These reactions have in common that they are *concerted*, and that they proceed through a *cyclic* transition state. The three main classes of pericyclic reaction are *electrocyclic* reactions involving ring closure in conjugated π-systems or its reverse, *sigmatropic* reactions in which a σ-bond migrates with respect to a π-framework, and *cycloadditions* and their reverse. R. B. Woodward and R. Hoffmann, in particular, used the concepts of orbital symmetry to predict which types of cyclic transition state are energetically feasible, and to predict the stereochemical consequences. The well-known

Woodward–Hoffmann rules embody the results, and use correlation diagrams extensively. Other formally-distinct (but theoretically-related) approaches to selection rules for pericyclic reactions include the use of *frontier orbitals* and the *aromatic transition state* concept, with the associated idea of *Hückel* and *Möbius* cyclic polyenes (the Möbius form has an odd number of twists, thus making the topology of the π-system that of a Möbius strip). This book is not the place to describe the theory of concerted reactions in any detail. The reader is directed to the Bibliography for some suggested texts that develop the subject more fully. We wish just to give an idea of how the approaches are extended to accommodate excited reactant species. Fortunately, the different approaches almost always lead to the same results (in both thermal and photochemical reactions). Each approach has its own merits in giving insight into particular types of process. We adopt the use of correlation diagrams, since that approach fits in with our treatment of the conservation of spin (and orbital) momentum.

We shall examine, as our example, the electrocyclization of a substituted buta-1,3-diene to a cyclobutene

$$(6.8a)$$

$$(6.8b)$$

The ring can be regarded as being closed by the formation of a σ-bond by the overlap of p-orbitals on carbon-1 and carbon-4 of the butadiene. Two modes of geometrical change must therefore be considered: *conrotatory*, as in reaction (6.8a), and *disrotatory*, as in reaction (6.8b). It is assumed that the presence of the substituents A, B, C, D does not appreciably distort the electronic orbitals from the symmetry that they would possess in unsubstituted buta-1,3-diene. However, the rotations that lead to the transition state and thence the products *do* reduce the symmetry of the system, and the symmetry element lost depends on which motion—conrotatory or disrotatory—is involved. The relevant symmetry elements possessed by buta-1,3-diene itself are indicated at the top of Fig. 6.1, while the lower part of the figure shows what happens as the transition states develop. For conrotatory cyclization, the vertical mirror plane is lost, while for disrotatory cyclization the axis of rotational symmetry disappears.

In order to understand whether conrotatory or disrotatory ring closures are 'allowed' in thermal or in photochemical reaction, we shall wish to construct a *state correlation diagram* showing which electronic states of the reactant correlate with which of the product. The first stage in constructing

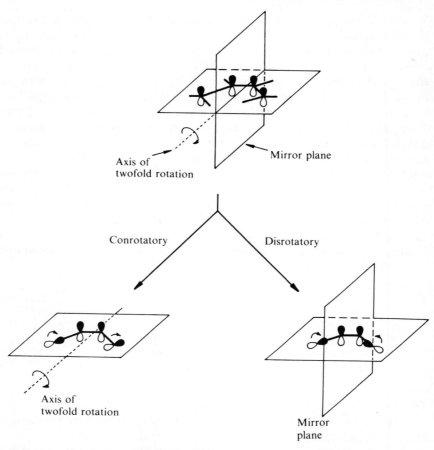

Fig. 6.1   The symmetry of butadiene and of the individual p-orbitals in conrotatory
and disrotatory ring closure.

the diagram is to find out the correlations for the individual orbitals. The
significant orbitals in butadiene are ψ_1, ψ_2, ψ_3, and ψ_4, while those in
cyclobutene are σ, π, π^*, and σ^* in ascending order. Figure 6.2 shows these
orbitals schematically, the positive lobes of the wavefunctions being shaded.
The levels are labelled 'S' or 'A' according to whether they are symmetric or
antisymmetric with respect to the remaining symmetry element in the conrotatory (two-fold axis) or disrotatory (mirror plane) case. The next stage is to
join up ('correlate') states of like symmetry in ascending order of energy, as
indicated by the dashed lines in the diagram. Ground-state butadiene is $\psi_1^2\psi_2^2$,
which can be seen from the figure to correlate with ground-state cyclobutene
($\sigma^2\pi^2$) in the conrotatory mode, but not in the disrotatory closure. We might

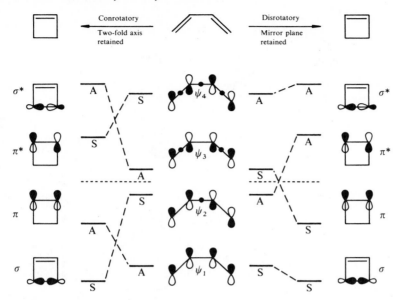

Fig. 6.2 Orbital correlation diagrams for the butadiene→cyclobutene reaction in the two ring-closure modes. The dots on the butadiene skeleton represent nodes in the wave function, so that the successively larger number of nodes (0, 1, 2, 3) shows the ordering in energy of ψ_1, ψ_2, ψ_3, ψ_4.

already conclude, therefore, that the *thermal* reaction is permitted for conrotatory cyclization, but forbidden in the disrotatory mode.

More insight may be gained by proceeding to the state correlation diagram. Consider first the disrotatory mode. The orbital diagram, Fig. 6.2, shows that ground-state, S_0, butadiene ($\psi_1^2\psi_2^2$) correlates with an excited singlet, S_m, of cyclobutene ($\sigma^2\pi^{*2}$). Similarly, ground-state cyclobutene ($\sigma^2\pi^2$) correlates with a highly excited state, S_n, of butadiene ($\psi_1^2\psi_3^2$). Finally, the first excited singlet, S_1, in butadiene ($\psi_1^2\psi_2\psi_3$) correlates with S_1 in cyclobutene ($\sigma^2\pi\pi^*$). These conclusions are summarized in Fig. 6.3(a). In this zero-order approximation, the point is re-emphasized that the ground-state reactant does not correlate with the ground-state product, and the reaction cannot proceed adiabatically. Constructing the equivalent state correlation diagram for the conrotatory electrocyclization shows immediately that S_0 of the reactant correlates with S_0 of the product, and that the reaction can therefore proceed adiabatically directly from the ground-state reactant.

A further refinement requires us to ask the question of whether we have allowed lines connecting states of like symmetry to cross, and thus contravene the 'non-crossing rule'. The symmetries of the states are determined by multiplying out the symmetries, A or S, of the individual electrons involved,

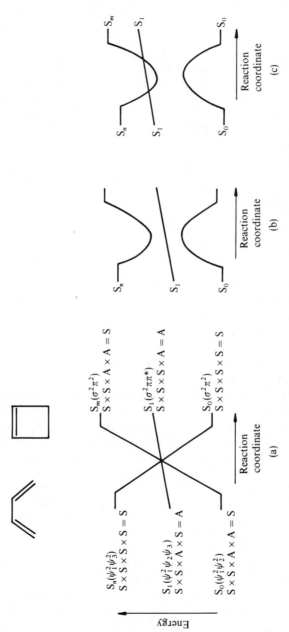

Fig. 6.3 State correlation diagrams for the disrotatory butadiene→cyclobutene reaction: (a) disregarding the rule that forbids crossing of states of similar symmetry; (b) avoiding the crossings; and (c) showing a more realistic form that takes account of the likely energy positions of S_1 and S_n states.

remembering that $S \times S = S$, $S \times A = A$, $A \times A = S$, as indicated on Fig. 6.3(a). Two 'S' lines *do* cross, so that Fig. 6.3(a) must be redrawn as Fig. 6.3(b). This state correlation diagram represents energy as a function of the reaction coordinate (a measure of the progress from reactants to products). There is clearly a sizeable energy barrier to reaction on the S_0 curve (surface). This first-order approximation thus shows that although the reaction can proceed adiabatically, on a continuous surface, it will do so with very low probability because of the activation barrier. It is therefore convenient to continue to regard the process as non-adiabatic (i.e. the terminology refers to the zero-order approximation: see also footnote on p. 103).

Figure 6.3 (a) or (b) shows that S_1 in butadiene correlates with S_1 in cyclobutene in the disrotatory mode, and gives a first indication that disrotatory cyclization might be allowed in a photochemically promoted reaction. Experiments show that photochemical disrotatory cyclization does, indeed, occur. However, the excited reactant does not pass over to the excited product, because S_1 in cyclobutene lies, in energy, over 200 kJ mol^{-1} above S_1 of butadiene. Rather, in this system, the course of events is illustrated by the more realistic representation of Fig. 6.3(c), in which the S_n curve effectively crosses the S_1 curve. Two radiationless transitions occur, first from S_1 to S_n, and then from S_n to S_0. The only energy deficit here is the rise between the initial S_1 and the crossing point on S_n; it is too small to afford a serious kinetic constraint. It may be noted that in certain other reactions, the excited-state reactant correlates with the ground-state product, and the reaction can then proceed directly.

Much experimental work shows that the stereochemical course of pericyclic reactions is in general opposite according to whether it is the ground or the excited state that reacts. For example, thermal cyclodimerization of substituted ethenes occurs predominantly by *cis–trans* addition, while photo-cyclodimerization yields mostly *cis–cis* addition products. The explanation for the opposing behaviour for thermal and photochemical reactions can be sought in terms of the correlation diagrams. The energy barrier that makes a ground-state reaction forbidden is a result of the correlation of the reactant and product ground states with upper excited states. An energy minimum *must* exist in the upper state of the same symmetry, as the derivation of Fig. 6.3(b) from Fig. 6.3(a) indicates. The high density of upper states means that the correlation diagrams generally resemble Fig. 6.3(c).

6.4 Intramolecular processes: isomerization and rearrangement

A wide variety of isomerizations and rearrangements can be induced photo-chemically. The use of *cis–trans* isomerization as an indicator of triplet energy transfer has already been referred to several times in Chapter 5. The occurrence of such isomerization may be used to establish the energy of triplet levels in donors and to determine the quantum yield of intersystem crossing $S_1 \leadsto T_1$ in donor molecules (p. 119). The availability of the rearrangement energy in suitable acceptors (e.g. *cis*-stilbene) has also been mentioned (p. 109) in connection with the activation energies of some endothermic energy transfer processes. A more complete discussion of these topics is given in the review by Wagner and Hammond (Bibliography to this chapter); here we shall discuss the *cis–trans* isomerization process itself, rather than pursue further the applications of the 'chemical' technique for elucidation of photophysical mechanism. We shall also consider briefly some examples of structural and valence isomerization brought about by absorption of light.

Ethene and its derivatives have lowest excited singlet and triplet states that are (π, π^*) in character: an electron is promoted from the highest filled bonding π orbital to the lowest antibonding π^* orbital. Both singlet and triplet (π, π^*) states are possible, the state of high multiplicity being of lower energy. As pointed out on p. 127, it can be shown that the (π, π^*) state is most stable if the molecule is twisted, from the planar ground-state configuration, through 90° about the double-bond axis. This perpendicular configuration minimizes the overlap between π and π^* orbitals, and both S_1 and T_1 states have their lowest energies for the 90° twist. It is apparent that if an alkene is excited to the (π, π^*) state, then it will tend to twist to the perpendicular configuration. Subsequent electronic energy degradation to the ground state will then require that the molecule become planar again, both *cis* and *trans* isomers being formed. The perpendicular form of the excited state is geometrically equivalent whether it is derived from a *cis* or *trans* ground-state molecule. It is the increased stability of the perpendicular configuration that sometimes makes the energy difference $S_0 - T_1$ obtained from absorption spectra incompatible with the energy for T_1 suggested by the rates of energy transfer (cf. p. 109). Phosphorescence from alkene triplets is not observed, and this is also probably a result of the relaxation of the triplet to a perpendicular form: the geometries of excited and ground states are so different that Franck–Condon restrictions effectively forbid emission to the lower vibrational levels of the ground state.

Geometrical photoisomerization can occur from *cis* to *trans* or from *trans* to *cis* isomers, so that on prolonged irradiation of either isomer, a photostationary state is set up. If the quantum yields for processes in each direction are similar, the concentrations of *cis* and *trans* isomers at the stationary state are determined mainly by the extinction coefficient of each isomer at the wavelength employed. For many simple alkenic systems, the higher extinction coefficient at longer wavelengths is possessed by the *trans* isomer, and if long wavelength light is used, the *cis* isomer will predominate at the photostationary state.

The photostationary concentrations in a *sensitized* isomerization might be expected to depend solely on the rate of conversion of the common triplet to the two ground-state isomers. With sensitizers whose triplet energy lies well above that of the alkene, this expectation is borne out. For example, the sensitized geometrical isomerization of stilbenes ($C_6H_5-CH=CH-C_6H_5$) gives a photostationary mixture containing roughly equal amounts of *cis* and *trans* isomers so long as the energy of the sensitizer is higher than the energy of either the *cis* (239 kJ mol^{-1}) or the *trans* (205 kJ mol^{-1}) triplet. The exact composition is equivalent to the branching ratio for decay of the common orthogonal triplet state to the isomeric ground states. Sensitizers whose triplet energy lies between that of *trans*- and *cis*-stilbene produce a selective isomerization of *trans*-stilbene, and the [*cis*]/[*trans*] ratio in the photostationary mixture reaches nearly 15 for sensitizer triplet energies of around 210 kJ mol^{-1}. Below this energy, neither *cis* nor *trans* linear triplets can be populated. However, so long as there is enough energy to excite the orthogonal triplet, isomerization can still occur, although the proportion of *cis* isomer in the stationary state rapidly decreases. Conjugated dienes, such as penta-1,3-diene, show the opposite behaviour to that of the stilbenes in the intermediate energy range, with the [*cis*]/[*trans*] ratio dropping below unity. Part of the explanation lies in the energy of the vertical *cis* triplet (239 kJ mol^{-1}) being lower than that of the *trans* triplet (247 kJ mol^{-1}) in the pentadiene, although there are other factors at work as well.

Photochemical valence and structural isomerizations are well known. The electrocyclic ring-closing reactions of dienes and trienes, discussed in Section 6.3, are typical of the valence isomerizations (e.g. reaction (6.8), p. 133). These electrocyclic processes require the *cis* conformation. An alternative cyclization yields a bicyclobutane

$$(6.9)$$

Since this product is also formed in the triplet sensitized reaction, it is possible that ring closure involves a diradical in a non-concerted process. Where a substituted diene is constrained to be *s-trans*, the bicyclobutane may even be the major product of cyclization.

The reactions of polyenes are often different for excited singlet and triplet states. Triplet-sensitized reaction usually leads to (cyclic) dimerization (Section 6.7), while the direct reaction leads to internal cyclization. For example, entirely different products are observed on direct and sensitized irradiation of myrcene

$$(6.10a)$$

$$(6.10b)$$

Since the products of reaction (6.10b) are probably formed via the triplet of the triene, it may be inferred that the singlet is responsible for the products of the direct photolysis. An example of a molecule that yields the *same* products on direct and sensitized photolysis is afforded by the cyclic diene cyclo-octa-1,3-diene

$$\text{(6.11)}$$

Aromatic compounds often cyclize to give polycyclic, non-aromatic products in the first instance, as exemplified by the photocyclization of *cis*-stilbene to dihydrophenanthrene

$$\text{(6.12)}$$

a process that accompanies direct *cis-* → *trans*-isomerization. Valence isomerization of benzene itself produces the highly reactive species benzvalene (and, at short wavelengths, bicyclohexadiene and fulvene as well)

$$\text{(6.13)}$$

Benzvalene   Bicyclo-   Fulvene
                                hexadiene

Benzvalene intermediates are responsible for a remarkable phototransposition of skeletal carbon atoms in substituted benzenes. 1,2-Dimethylbenzene (*o*-xylene) is transformed first to the 1,3-isomer and subsequently to the 1,4 isomer as a result of transposition of adjacent ring carbons

$$\text{(6.14)}$$

Structural photoisomerizations, with ring rearrangment, also occur with cyclic conjugated enones and dienones. Dialkylcyclohexenones isomerize with migration of the bond between carbon-4 and carbon-5.

$$\text{(6.15)}$$

The reaction proceeds via the triplet (π, π^*) state. Apparently, the isomerization is a concerted and stereospecific reaction; such behaviour, as suggested earlier, is unusual for a triplet. Cyclic dienones isomerize in a similar way, and the second double bond in the ring must play an important part in the mechanism, because higher quantum efficiencies are obtained than with the simpler enones. For example, with 4,4-diphenylcyclohexa-2,5-dienone the quantum efficiency approaches unity. The excited state involved is the triplet (π, π^*). Diarylcyclohexenones isomerize differently, with the substituent on carbon-4 migrating

$$(6.16)$$

Reaction (6.16) is similar to another important photoisomerization known as the *di-π-methane* rearrangement in which 1,4-dienes and 3-phenylalkene systems yield vinylcyclopropanes

$$(6.17)$$

The mechanism is formally a 1,2-shift with ring closure,

$$(6.18)$$

although diradical intermediates have not been detected directly. The reaction is usually promoted via the excited singlet, but, in spite of the observed stereospecificity, it does not seem to be a concerted process.

Finally, in this discussion of photoisomerization, we must give an example of a photochemical sigmatropic shift. As described in Section 6.3, sigmatropic shifts are concerted pericyclic reactions in which a σ-bond migrates with respect to a system of π-electrons, with consequent switching of double and single bonds in the system. The orbital symmetry (Woodward-Hoffman) rules show that for photochemical suprafacial shifts there must be $4n$ electrons involved, where n is an integer (and $4n + 2$ electrons in an antarafacial shift). As a result, in the more commonly observed suprafacial shifts, atoms are transferred from the initial position to the third or seventh carbon atom in the system: four and eight electrons are involved, respectively. Examples of these so-called (1, 3) and (1, 7) photochemical sigmatropic shifts are given in eqns. (6.19) and (6.20).

$$\text{(structure)} \xrightarrow{h\nu} \text{(structure)} \tag{6.19}$$

$$\text{(structure)} \xrightarrow{h\nu} \text{(structure)} \tag{6.20}$$

6.5   Intramolecular processes: hydrogen abstraction

Hydrogen abstraction is one of the most important intramolecular reactions of excited species. The process is typical of molecules possessing lowest excited states (n, π*) in character (e.g. aldehydes and ketones): indeed those 'abnormal' carbonyl compounds whose lowest excited levels are (π, π*) states (cf. Section 6.2, p. 127) undergo neither intramolecular nor intermolecular H-abstraction, except with very good donors such as amines that provide hydrogen by an indirect route. The intramolecular abstraction of H is of particular importance in the photochemistry of most carbonyl compounds, since it is part of the sequence of events leading to the 'Norrish Type II' fragmentation (Section 3.6, p. 53). Equation (3.18) shows a six-membered transition state in the Type II fission of a ketone: this cyclic intermediate favours intramolecular H-abstraction over intermolecular abstraction from the solvent, and the individual steps of eqn (3.18) may be written more fully as

$$CH_3CH_2CH_2COCH_3 \;+\; h\nu \;\longrightarrow\; \begin{array}{c} H \quad O \\ H_2C \qquad C-CH_3 \\ H_2C-CH_2 \end{array}$$

intramolecular
H-abstraction

$$\begin{array}{c} OH \\ H_2C\cdot \quad \cdot C-CH_3 \\ H_2C-CH_2 \end{array} \longrightarrow \begin{array}{c} CH_2=COH \\ CH_3 \\ + \\ CH_2=CH_2 \end{array} \longrightarrow CH_3COCH_3 \tag{6.21}$$

Diradicals have been detected (e.g. by flash spectroscopy), and the enols produced by cleavage have also been investigated spectroscopically at low temperatures. Aryl ketones normally react *via* a triplet (n, π*) state, and the reactions are not stereoselective. Dialkyl ketones, on the other hand, react through both singlet and triplet states, and the singlet reaction is highly stereoselective.

In addition to the fragmentation reaction of the hydroxy diradical formed by intramolecular H-abstraction, there is a ring-closure path leading to cyclobutanol formation:

$$
\underset{\underset{\displaystyle \text{R—C—CH}_2\text{CH}_2\text{CH}_2}{|}}{\overset{\displaystyle \text{OH}}{|}} \longrightarrow \underset{\underset{\displaystyle \text{CH}_2\text{—CH}_2}{|}}{\overset{\displaystyle \text{OH}}{\underset{|}{\text{R—C——CH}_2}}} \tag{6.22}
$$

In certain cases, the photochemical cyclobutanol formation appears to be stereospecific, e.g.

$$
(CH_3)_2CH-\overset{O}{\overset{\|}{C}}\overset{CH_2-CH_2}{\underset{CH_2=CH}{\diagdown}}\overset{CH_3}{\underset{H}{C}} \xrightarrow{h\nu} (CH_3)_2CH-\overset{OH}{C}\overset{CH_2-CH_2}{\diagdown}\overset{CH_3}{\underset{CH=CH_2}{C}}
$$
$$\tag{6.23}$$

The retention of configuration suggests that ring closure is so rapid that rotation about the C–C bond does not occur (although a concerted process may also operate for the singlet-state reactions). Diradicals other than 1,4-diradicals do not undergo the cleavage reaction, so that photocyclization becomes relatively more important in systems where H-abstraction cannot occur via the six-membered transition state, i.e. where the γ-position of the carbonyl compound does not carry a hydrogen atom. Where both cleavage and cyclization can occur, the branching ratio may depend on experimental parameters such as temperature and pressure. Higher yields of the cyclization product can also be obtained by using ordered arrays of molecules (e.g. in micelles). It seems likely that the orbitals of the diradical electrons and of the bond to be broken must be aligned parallel for efficient cleavage to occur.

6.6 Intermolecular processes: hydrogen abstraction

Excited (n, π^*) states, especially those of aromatic carbonyl compounds, not only undergo the intramolecular reduction reactions described in the last section, but can also abstract hydrogen in intermolecular processes. We referred in Section 6.2 to the photoreduction of benzophenone in the presence of a suitable hydrogen donor:

$$
C_6H_5COC_6H_5 + RH \xrightarrow{h\nu} C_6H_5C(OH)C_6H_5 + R \tag{6.24}
$$

In a good hydrogen-donating solvent, such as ethanol, the quantum yield for benzophenone removal is near unity, although in solvents for which the activation energy for H-abstraction is high, the quantum yield may be much smaller. With substrates of low ionization potential, such as amines or unsaturated hydrocarbons, the favoured process may be charge transfer or electron abstraction to yield a ketyl radical anion. Under suitable conditions, quantum yields around *two* are observed for removal of benzophenone in propan-2-ol solutions. The radical produced from the solvent is itself capable of reducing a molecule of benzophenone to its ketyl radical, and two molecules of benzophenone are removed for each quantum of light absorbed:

$$(C_6H_5)_2CO + (CH_3)_2CHOH \xrightarrow{h\nu} (C_6H_5)_2COH + (CH_3)_2COH \qquad (6.25)$$

$$(CH_3)_2COH + (C_6H_5)_2CO \longrightarrow (CH_3)_2C=O + (C_6H_5)_2COH \qquad (6.26)$$

The ketyl radicals formed can participate in secondary reactions, which include dimerization to form a pinacol:

$$2C_6H_5C(OH)C_6H_5 \longrightarrow C_6H_5\!-\!\overset{\overset{\displaystyle OH}{|}}{\underset{\underset{\displaystyle C_6H_5}{|}}{C}}\!-\!\overset{\overset{\displaystyle OH}{|}}{\underset{\underset{\displaystyle C_6H_5}{|}}{C}}\!-\!C_6H_5 \qquad (6.27)$$

or further hydrogen abstraction (in alkaline solution) to form benzhydrol, $(C_6H_5)_2CHOH$. Photopinacolization has been recognized since the beginning of the century, when it was found that good yields of benzpinacol were produced by the action of sunlight on solutions of benzophenone.

There is direct optical and ESR spectroscopic evidence for the existence of ketyl radicals and ketyl radical anions in the photoreduction of ketones. Both singlet and triplet (n, π^*) excited states of ketones can undergo allowed in-plane hydrogen and electron abstraction reactions; normally, the triplet is the major contributor to photoreduction, although the singlet may occasionally be more reactive with some substrates because of the energetics of abstraction.

The efficiency of photoreduction depends not only on the nature of the solvent, but also on the structure of the ketone. Substitution of aryl ketones in the 2-position makes possible the *intra*molecular hydrogen abstraction discussed in the last section. A six-membered cyclic transition state is available, and photopinacolization is replaced by the formation of an unsaturated compound.

$$(6.28)$$

Steric hindrance is not the main factor in reducing the efficiency of inter-molecular hydrogen abstraction, because where the 2-substituent has no available hydrogens (i.e. where the six-membered ring cannot be formed), pinacols are formed quite efficiently.

Table 6.1 *Rate constants for reduction of ketone triplets (Data from Turro, N. J. (1978). Modern molecular photochemistry, Table 10.2, p. 378. Benjamin/Cummings, Menlo Park, CA)*

Ketone	Donor	k_r (dm^3 mol^{-1} s^{-1})
Benzophenone	Propan-2-ol	1×10^6
4-CF$_3$-Benzophenone	Propan-2-ol	2×10^6
4-CH$_3$-Benzophenone	Propan-2-ol	1×10^5
4-C$_6$H$_5$-Benzophenone	Propan-2-ol	1×10^4
4,4'-Bis(dimethylamino) benzophenone	Cyclohexane	$<2 \times 10^3$

Substitution in the 4-position of benzophenone seems to influence the efficiency of photoreduction by altering the nature of the excited state. Table 6.1 shows rate constants for reduction of the triplets of benzophenone and some of its derivatives. We have referred already to the spectroscopy of these 'abnormal' ketones. The (n, π^*) and (π, π^*) configurations are much closer together in energy for aryl ketones than for alkyl ketones, and, in the case of some substituted derivatives, the lowest state can be the (π, π^*). For example, the phosphorescence lifetime of 4-phenylbenzophenone is about 50 times longer than that of 'normal' benzophenones, suggesting that the lowest triplet is (π, π^*) in character. The structure of the emission spectrum and EPR studies both confirm this conclusion. In the (π, π^*) state, the excited carbonyl oxygen is not as electron deficient as in the (n, π^*) state, and the excitation energy is also partly delocalized into the π-system so that it is not all available for overcoming activation energies. The consequence is that the (π, π^*) state is likely to be much less reactive than the (n, π^*) state, and photoreduction of the 4-phenyl ketone will be correspondingly inefficient. The 4-methyl ketone probably has a triplet with mixed character, and the rate of reduction is intermediate between benzophenone and 4-phenylbenzophenone. If electron-releasing groups are present as substituents, such as in the aminobenzophen-ones, charge-transfer states such as

become the lowest triplets, at least when stabilized in propan-2-ol solution. These states are even less reactive than the (π, π^*) states. In non-polar solution the lack of solvation leaves the ordinary (n, π^*) state as the lowest triplet, and relatively high quantum yields for photoreduction are observed (e.g. 0.9 for 4-hydroxybenzophenone in cyclohexane, as against 0.02 in propan-2-ol).

6.7   Intermolecular processes: addition reactions

A wide variety of photoaddition reactions is known, and both homoaddition and heteroaddition reactions can occur. Alkenes can undergo photochemical electrophilic addition, for example with water, alcohols, and carboxylic acids. Photochemical additions involving excitation of an aromatic compound are also known, as in the reaction with an amine

$$\text{(6.29)}$$

Many photoaddition reactions lead to cyclic products, and we shall examine briefly some important examples of these cycloadditions.

Two σ-bonds are formed in a cycloaddition reaction to give a new ring. Common classes of cycloaddition include the $(2+2)$ addition of two alkenes to give a cyclobutane

$$\text{(6.30)}$$

and the $(4+2)$ addition of a conjugated diene and an alkene to give a cyclohexene

$$\text{(6.31)}$$

Concerted $(2+2)$ photochemical cycloadditions are allowed on orbital symmetry grounds if they occur in a suprafacial–suprafacial manner. The dimerization of pure but-2-ene is an example of such a concerted cycloaddition. Irradiation of either the *cis* or the *trans* isomer on its own gives rise to a pair of isomers of 1,2,3,4-tetramethylcyclobutane with the expected stereochemistry. More often, however, the cycloaddition reactions of alkenes are not concerted, but are two-step diradical reactions. We have already seen in reaction (6.10b) how triplet sensitizers can promote dimerization. Open-chain dienes, such as buta-1,3-diene, form cyclobutane derivatives and, in addition, 4-vinylcyclohexene. At high sensitizer triplet energy (e.g. acetophenone, $E_T \sim 310 \text{ kJ mol}^{-1}$), the cyclobutane derivatives are the major products, but with sensitizers whose energy lies below about 251 kJ mol^{-1}, the fraction of cyclobutanes drops off relative to the vinylcyclohexene. The $S_0 \rightarrow T_1$ energy for the *s-trans-* rotamer of butadiene is itself about 251 kJ mol^{-1}, while that for *s-cis*-butadiene is probably lower (possibly about 226 kJ mol^{-1}, an energy below which ketones do not efficiently sensitize the dimerization to any of the products). The different yields of the dimers may, therefore, be a result of the existence of non-interconvertible, stereoisomeric triplet states of open-chain dienes.

With cyclic conjugated dienes that are rigidly held in the *s-cis* configuration, the relative yields of the isomeric products do not depend on the nature of the sensitizer. For example, in the dimerization of cyclohexa-1,3-diene, the isomers

are formed in relative yields of about 60, 19, and 21% for sensitizers whose triplet energy ranges between 222 and 289 kJ mol^{-1}.

Benzene and its derivatives can undergo cycloaddition across the 1, 2-, 1, 3-, or 1, 4-positions. The 1, 3-addition is the main mode of reaction with simple alkenes,

$$(6.32)$$

but it seems to be restricted to double bonds bearing only alkyl substituents. Addition is stereospecific, and involves singlet excited benzene. 1, 2-Addition predominates if the electron donor and electron acceptor properties of the alkene and the aromatic species are markedly different. Thus alkenes that are electron-rich (e.g. 1,1'-dimethoxyethene) or electron-deficient (e.g. acrylonitrile) are particularly likely to add across the 1, 2-positions to produce primary adducts of the bicyclo[4.2.0]-octa-2,4-diene type. Equation (6.33) illustrates the generalized 1, 2-addition.

$$(6.33)$$

If the process is concerted, as the stereospecificity suggests, then the orbital symmetry constraints must be relaxed for some reason, because the reaction is forbidden for excited singlet benzene and ground-state alkene. One suggestion is that the addition occurs in an exciplex. The 1,4-cycloaddition is the least frequently encountered of the processes, but is again stereospecific and probably also occurs through an exciplex interaction.

Cyclobutane derivatives are formed by the photoaddition of alkenes to the unsaturated bond in α,β-unsaturated ketones. For example, cyclobutane derivatives are the major products of irradiation of cyclohex-2-enone in the presence of 2-methylpropene

$$(6.34)$$

$$(35\%) \qquad (16\%)$$

The reactive state is in most cases a triplet, and this observation may explain why cyclic enones undergo photocyclic addition more readily than their acyclic analogues. The reduced flexibility of the cyclic compounds inhibits intersystem crossing to the ground state, which rapidly deactivates the twisted triplet states of acyclic enones. The reactions are not stereospecific, but some orientations in the products are preferred, possibly because of the geometry of an exciplex formed as the first step in the interaction of the excited-state enone and ground-state alkene.

Excited aldehydes and ketones may take part in reactions that involve the addition of the excited carbonyl group itself to suitable alkenes. The product is an oxetane. For example, irradiation of benzophenone in the presence of 2-methylpropene leads to the formation of an oxetane in relatively high yield

$$(6.35)$$

The triplet (n, π^*) state of the carbonyl compound appears to attack the ground state of the alkene to yield a 1,4-diradical, which then cyclizes. The preferred orientation of the addition is determined by the stability of the possible diradicals, as in reaction (6.35), where the alternative oxetane isomer (with the CH_3 groups adjacent to the oxygen) accounts for $<10\%$ of the product. Those 'abnormal' ketones that do not undergo efficient photoreduction (cf. Section 6.6) are also inefficient in the oxetane reaction. Further, if the *alkene* has a triplet energy below that of the carbonyl compound, energy transfer from the carbonyl group takes place to the virtual exclusion of oxetane formation. For alkyl ketones, oxetanes can be formed *via* the first excited singlet (n, π^*) state.

Photochemical addition reactions of oxygen are thought to be important in many photosensitized oxidations of unsaturated compounds. The biological implications photosensitized oxidation have been recognized since 1900, when it was discovered that microorganisms can be killed in the presence of oxygen and sensitizing dyes. The pathological effects of photooxidation of cell constituents include cell damage, induction of mutations or cancer, and death. Recent investigations of photosensitized oxidation have led to a better understanding of the chemical processes, and the results are now finding application in the biological field. It seems appropriate to end the present chapter with a description of these highly significant photooxidation reactions.

Almost all photosensitized oxidations proceed via a triplet state of the sensitizer, presumably because this state has a much greater lifetime than the excited singlets. The first process involving the sensitizer triplet may be reaction either with the substrate or with the oxygen. For many dye triplets, the efficiency of reaction with oxygen is so great that this process predominates at all but the lowest oxygen concentrations. Whether the presence of oxygen merely inhibits the reaction with substrate, or whether the products of the primary reaction with O_2 take part in further reactions leading to oxidation of substrate, depends on the chemical nature of the substrate. A very efficient sensitized oxidation is observed with alkenes, dienes, dienoid heterocycles, and polycyclic aromatic compounds, and it is with these substances as oxidizible reactant that we are concerned. The first oxidation products are often peroxides or hydroperoxides, and they may subsequently take part in secondary oxidation steps.

As typical examples, we may consider the products formed on irradiation of dimethylfuran and of tetramethylethene in the presence of oxygen and a triplet sensitizer; in the first case, an endoperoxide (in fact an ozonide) is formed, while the product of the alkene oxidation is a hydroperoxide:

$$\text{H}_3\text{C}\!-\!\!\overset{\displaystyle}{\underset{\text{O}}{\boxed{}}}\!\!-\!\text{CH}_3 \;+\; \text{O}_2 \;\xrightarrow{h\nu,\ \text{sens}}\; \text{H}_3\text{C}\!-\!\!\overset{\displaystyle}{\underset{\underset{\text{O}-\!\!-\!\text{O}}{\text{O}}}{\boxed{}}}\!\!-\!\text{CH}_3 \qquad (6.36)$$

$$\begin{array}{c}\text{H}_3\text{C}\\ \text{H}_3\text{C}\end{array}\!\!\!\!>\!\!\text{C}\!=\!\text{C}\!<\!\!\!\!\begin{array}{c}\text{CH}_3\\ \text{CH}_3\end{array} \;+\; \text{O}_2 \;\xrightarrow{h\nu,\ \text{sens}}\; \begin{array}{c}\text{H}_3\text{C}\\ \text{H}_3\text{C}\end{array}\!\!\!\!>\!\!\underset{\text{ÓOH}}{\text{C}}\!-\!\text{C}\!<\!\!\!\!\begin{array}{c}\text{CH}_2\\ \text{CH}_3\end{array} \qquad (6.37)$$

Alkenes with electron-rich double bonds, or those that do not possess an allylic hydrogen, can undergo a $(2+2)$ cycloaddition to form a dioxetane

$$C_2H_5O-CH{=}CH-OC_2H_5 \;+\; O_2 \quad \xrightarrow{hv,\ sens} \quad C_2H_5O\underset{H}{\overset{O-O}{C}}{-}\underset{H}{\overset{\;}{C}}OC_2H_5 \qquad (6.38)$$

Dioxetanes can be quite stable, but they undergo chemiluminescent (see Section 4.7) thermal decomposition to produce two carbonyl fragments, one of which is electronically excited.

The kinetic data are consistent with the quantitative formation of some intermediate from the sensitizer triplet at oxygen concentrations above about 10^{-5} mol dm^{-3}. The intermediate is itself trapped quantitatively by good acceptors, although it fails to react at all with many compounds (e.g. alcohols). There are two possible interpretations of these facts: (i) that the intermediate is a sensitizer–oxygen complex, and (ii) that the intermediate is electronically excited oxygen formed by energy transfer from sensitizer to oxygen. These two paths probably both involve the formation of an adduct (an 'exciplex') between excited sensitizer and oxygen, but only in the first mechanism is the exciplex stable:

$$\text{(i)} \quad {}^3\text{Sens} + O_2 \rightarrow \text{Sens--O--O} \qquad (6.39)$$

$$\text{(ii)} \quad {}^3\text{Sens} + O_2 \rightarrow \text{Sens} + O_2^{\dagger} \qquad (6.40)$$

where ^{3}Sens represents the triplet state of the sensitizer.

Overwhelming evidence points to the occurrence of the energy transfer process in many sensitized photo-oxidations. Since the ground state of oxygen is a triplet, it is necessary for $O_2^{\dagger}$ to be a singlet state for reaction (6.40) to be spin-conserved. The lowest-lying excited state of oxygen is, in fact, a singlet ($^1\Delta_g$) and possesses an excitation energy of 92 kJ mol^{-1}, so that it can easily be excited by energy transfer from the triplet states of most dyes.

Direct studies of the reactions of $O_2(^1\Delta_g)$ have indicated that it is the intermediate involved in sensitized photo-oxidation. The excited species may be produced in a number of ways: for example, in the reaction of sodium hypochlorite with hydrogen peroxide (cf. p. 93) or by the action of a microwave discharge on molecular oxygen in the gas phase. For a wide variety of acceptors that yield more than one oxidation product, the product distributions from the reaction with $O_2(^1\Delta_g)$ and from the photo-oxidation are identical, and there are no detectable differences in stereoselectivity. If the photo-oxidation involved a bulky sensitizer complex with O_2 at the transition state, a quite different stereoselectivity and product distribution might be expected. Further, the ratio of rate constants for decay and for reaction with acceptor are identical for $O_2(^1\Delta_g)$ and for the intermediate in photo-oxidation. It has also been shown explicitly that $O_2(^1\Delta_g)$ can be formed in reaction (6.40); the emission band at 1270 nm, from the $O_2(^1\Delta_g \rightarrow {}^3\Sigma_g^-)$ transition, is observed on irradiation in the gas phase of mixtures of oxygen with suitable triplet donors (e.g. benzaldehyde). This piece of evidence adds considerable weight to the arguments favouring $O_2(^1\Delta_g)$ as the intermediate in the sensitized photo-oxidations.

The elucidation of the mechanism of sensitized photo-oxidation has made possible several fruitful speculations with regard to photobiology. As an example, we will consider the protective action of carotenoids in biological systems. Carotenoids apparently protect photosynthetic organisms against the lethal effects of their own chlorophyll (see p. 201), which is an excellent sensitizer of photo-oxidation. It has been shown that β-carotene is an extremely efficient quencher of singlet oxygen, and it can also inhibit sensitized photo-oxidations. For example, β-carotene, at a concentration of $10^{-4}\,\text{mol dm}^{-3}$, inhibits 95% of the methylene-blue-sensitized oxidation of 2-methyl-2-pentene ($10^{-1}\,\text{mol dm}^{-3}$); under the conditions of the experiment ($[O_2] = 10^{-2}\,\text{mol dm}^{-3}$), virtually no triplet methylene blue is quenched by the β-carotene, so that the inhibition derives from the quenching of $O_2(^1\Delta_g)$. Carotene does not appear to be consumed in the reaction, which suggests that the quenching of $O_2(^1\Delta_g)$ may involve excitation of triplet carotene by energy transfer:

$$O_2^{\ddagger}(^1\Delta_g) + \beta\text{-carotene} \rightarrow O_2(^3\Sigma_g^-) + {}^3\beta\text{-carotene} \qquad (6.41)$$

Thus, the interesting speculation may be made that carotenoids serve a double function in photosynthetic organisms: first, that they remove 'toxic' singlet oxygen, and, secondly, that they can store the energy that O_2 receives from chlorophyll, which would otherwise be lost.

Bibliography

Coyle, J. D. (1986). *Introduction to organic photochemistry*. John Wiley, Chichester and New York.

Turro, N. J. (1978). *Modern molecular photochemistry*, Chapters 10–14. Benjamin/Cummings, Menlo Park, CA.

Calvert, J. G. and Pitts, J. N., Jr (1966). *Photochemistry*, Chapter 5. John Wiley, Chichester and New York.

Barltrop, J. A. and Coyle, J. D. (1975). *Excited states in organic chemistry*, Chapters 6–11. John Wiley, Chichester and New York.

Horspool, W. M. (1976). *Aspects of organic photochemistry*. Academic Press, New York.

Cowan, D. O. and Drisko, R. L. (1976). *Elements of organic photochemistry*. Plenum Press, New York.

Horspool, W. M. (ed.) (1984). *Synthetic organic photochemistry*. Plenum Press, New York.

Padwa, A. (ed.) (1979–1987). *Organic photochemistry*, Vols 4–9. Marcel Dekker, New York.

Special topics
Section 6.2

Julliard, M. and Chanon, M. (1982). Redox properties of photoexcited states. *Chem. Br.* **18**, 558.

Wagner, P. J. (1983). Conformational flexibility and photochemistry. *Acc. Chem. Res.* **16**, 461.

Wagner, P. J. and Hammond, G. S. (1967). Reactions and properties of organic molecules in their triplet states. *Adv. Photochem.* **5**, 21.

Wagner, P. J. (1976). Chemistry of excited triplet organic carbonyl compounds. *Top. Curr. Chem.* **66**, 1.

Section 6.3

Woodward, R. B. and Hoffmann, R. (1970). *The conservation of orbital symmetry.* Verlag Chemie International, Deerfield Beach, FL.

Zimmerman, H. E. (1971). The Möbius–Hückel concept in organic chemistry. Application to organic molecules and reactions. *Acc. Chem. Res.* **4**, 272.

Gilchrist, T. L. and Storr, R. C. (1979). *Organic reactions and orbital symmetry.* Cambridge University Press, Cambridge.

Pearson, R. G. (1976). *Symmetry rules for chemical reactions.* John Wiley, Chichester and New York.

Barltrop, J. A. and Coyle, J. D. (1978). *Principles of photochemistry*, Chapter 6. John Wiley, Chichester and New York.

Section 6.4

De Mayo, P. (ed.) (1980). *Rearrangements in ground and excited states*, Vol. 3. Academic Press, New York.

Mallory, F. B. and Mallory, C. W. (1984). *Organic reactions*, Vol. 30. John Wiley, Chichester and New York.

Schuster, D. I. (1978). Mechanisms of the photochemical transformations of cross-conjugated cyclohexadienones. *Acc. Chem. Res.* **11**, 65.

Phillips, D., Lemaire, J., Burton, C. S., and Noyes, W. A., Jr (1968). Isomerization as a route for radiationless transitions. *Adv. Photochem.* **5**, 329.

Chapman, O. L. (1963). Photochemical rearrangements of organic molecules. *Adv. Photochem.* **1**, 323.

Section 6.5

Turro, N. J., Dalton, J. C., Davies, K., Farrington, G., Hautala, R., Morton, D., Niemczyk, M., and Schore, N. (1972). Molecular photochemistry of alkanones. *Acc. Chem. Res.* **5**, 92.

Wagner, P. J. (1971). Type II photoelimination and photocyclization of ketones. *Acc. Chem. Res.* **4**, 168

Section 6.6

Scaiano, J. C. (1973). Intermolecular photoreductions of ketones. *J. Photochem.* **2**, 81.

Section 6.7

Eaton, P. E. (1968). Photochemical reactions of simple alicyclic enones. *Acc. Chem. Res.* **1**, 50.

Arnold, D. R. (1968). The photocycloaddition of carbonyl compounds to unsaturated systems: the synthesis of oxetanes. *Adv. Photochem.* **6**, 301.

McCullough, J. J. (1987). Photoadditions of aromatic compounds. *Chem. Rev.* **87**, 811.

Foote, C. S. (1968). Photosensitized oxygenations and the role of singlet oxygen. *Acc. Chem. Res.* **1**, 104.

Frimer, A. A. (ed.) (1985). *Singlet oxygen*, Vols. 1–4. CRC Publishing, Boca Raton, FL.

7

Techniques in photochemistry

7.1 Introduction

A proper understanding of the principles of photochemistry requires some appreciation of the methods used in the various studies; the present chapter provides a short description of the more common techniques. Detailed discussion of apparatus is, however, purposely excluded; several articles provide this more specific information (see Bibliography).

Photochemical processes may lead to chemical change; the nature of the products, and the rates of their formation, may be determined by standard chemical techniques which need not be treated here. We are more concerned with those parts of the experimental technique that involve light. Measurements of absorbed (and, sometimes, emitted) light intensities are essential to determinations of quantum yields, which are themselves needed in any assessment of the efficiency of primary photochemical processes. Quantum yields may be determined by the use of 'classical' techniques, i.e. under steady illumination. The kinetic behaviour of reaction systems under continuous illumination is often consistent with the presence of reactive intermediates at their stationary-state concentrations. Further kinetic data (individual rate constants, for example) are obtained from experiments performed with non-stationary conditions: this point has already been illustrated for photolysis (p. 14) and fluorescence (p. 74). Photochemical processes are ideally suited to non-stationary investigations, since the illumination may be started or stopped suddenly by use of a flash of light or by use of a mechanical shutter: it is often impossible to start or stop thermal reactions in the same way (although shock waves may, of course, be used to cause rapid heating in gaseous systems). In this chapter, we start by discussing the light sources that are used in photochemical experiments, and then provide a brief summary of classical photochemical techniques. We then go on to describe how the reactive intermediates in photochemical processes may be identified and investigated, and show how time-resolved experiments are used to obtain detailed kinetic and mechanistic information. We end with a brief indication of the methods being used in 'high resolution photochemistry' to yield an understanding of the detailed dynamics of photochemical interactions.

7.2 Light sources

Ideally, a photochemical experiment employs monochromatic light, since the nature of the primary processes, and their quantum efficiencies, may be wavelength-dependent. The use of monochromatic radiation also simplifies the measurement of absolute light intensities. Since most light-sources, except for lasers, are polychromatic, some technique must therefore be used to isolate a narrow wavelength band. Grating or prism monochromators are well suited to this purpose, although the light intensities available from them may not be sufficient for certain experiments. It is, perhaps, more usual to employ one or more 'colour filters': these may be liquid solutions or glasses of substances that strongly absorb light of unwanted wavelengths. Interference filters, which depend on the interference effects in thin films (akin to those giving rise to colours in soap bubbles) are of great value in photochemistry, since they may be constructed to have any desired transmission characteristics.

7.2.1 *Incandescent filament lamps*

The incandescent tungsten-filament lamp gives radiation whose continuous spectrum approximates to that of a black body, and, while ordinary lamps are often adequate for the production of visible radiation, extremely high operating temperatures are required to obtain significant intensities in the ultraviolet. The introduction of a little iodine into the lamp permits these high filament temperatures to be used without breakdown of the lamp: 'quartz-halogen' lamps (possessing a quartz bulb[†]) provide a useful, spectrally continuous source of ultraviolet down to $\lambda \sim 200$ nm.

7.2.2 *Discharge lamps*

Although there are purposes for which the continuum from incandescent lamps is useful, much higher intensities of near-monochromatic radiation are obtained by filtering radiation from lamps that emit most of their energy in a small number of narrow bands or lines. Several types of gas-discharge lamp may be employed for this purpose; they may contain inert gases or vapours of volatile (often metallic) elements that produce the appropriate atomic emission lines. At low pressures, most of the emitted energy may be concentrated

[†]Ordinary window glass (1 mm thick) transmits about 50% of the incident radiation at $\lambda \sim 320$ nm, while quartz of the same thickness has a transmission of 50% at $\lambda \sim 175$ nm (depending on the purity). All optical components (lamps, windows, lenses, reaction cells, etc.) must obviously be made from materials that are transparent at the wavelengths of interest. Approximate wavelengths at which the transmittance of 5-mm thick samples is 50% are, for various substances: window glass, 350 nm; Pyrex, 330 nm; crystal quartz, 185 nm; ultrapure quartz, 165 nm; CaF_2, 135 nm; LiF, 107 nm. Thinner samples provide useful transmission at significantly shorter wavelengths. Oxygen absorbs at wavelengths less than about 200 nm, and systems for study at $\lambda < 200$ nm must be air-free; at shorter wavelengths it is usual to employ evacuated systems (hence the term 'vacuum ultraviolet' for the spectral region with $\lambda < 200$ nm).

156   *Techniques in photochemistry*

in the 'resonance' lines (those due to transitions from the first excited state to the ground state), and essentially monochromatic light can be obtained without the use of filters: typical of such lamps are those containing low-pressure Xe ($\lambda = 147.0$ nm) or Hg ($\lambda = 184.9$ nm, 253.7 nm, cf. p. 29): in the latter case, a little inert gas is usually present, but contributes little to the actual emitted radiation. At higher pressures, created in discharges through metal vapours by operating the lamp at high temperature, more emission lines appear, and they show pressure broadening; the resonance line itself is often *reversed* as a result of absorption by cooler metal vapour near the walls of the lamp. Mercury discharge lamps are very frequently used in photochemical experiments, and Table 7.1 shows the intensities of the main lines from typical low-pressure (intensities relative to $\lambda = 253.7$ nm) and medium-pressure (intensities relative to $\lambda = 365.0$ nm) sources.

Table 7.1 *Relative intensities of some lines from mercury discharge lamps (Adapted from Calvert, J. G. and Pitts, J. N., Jr (1966).* Photochemistry, *Table 7.2, p. 696. Wiley, New York)*

	Relative intensity (energy units)	
λ (nm)	Low-pressure	Medium-pressure
579.0 (yellow)	10.1	76.5
546.1 (green)	0.88	93.0
435.8 (blue)	1.00	77.5
404.7 (violet)	0.39	42.2
365.0 (UV)	0.54	100.0
334.1	0.03	9.3
313.0	0.60	49.9
303.0	0.06	23.9
296.7	0.20	16.6
289.4	0.04	6.0
265.3	0.05	15.3
253.7	100	reversed

Discharge lamps may be designed for pulsed ('flash') operation. Xenon flash-lamps related to those used in photography are typical of such devices, and they find several photochemical applications, especially in flash photolysis (see Section 7.5) and as optical 'pumps' for lasers (see next section and Section 5.7). A xenon flash-lamp normally consists of a quartz tube filled with xenon to a pressure of a few hundred mmHg. It possesses two main electrodes connected to a charged capacitor that stores electrical energy. Ionization in the xenon is initiated by feeding a high-voltage trigger pulse to a third electrode, and a discharge expands to fill the tube. A fraction of the electrical energy is converted to light, which in the visible and ultraviolet regions roughly approximates to a continuum equivalent to black-body emission at a temperature of 7000–8000 K. There is thus useful intensity in the ultraviolet region right down to the cut-off of the quartz tube envelope. Pulse durations are normally of the order of microseconds, depending on the exact composition and pressure of the gas-fill and the electrical circuitry. Shorter flashes can be obtained, but there is usually a trade-off between the energy that can be dissipated and the flash duration.

7.2.3 Lasers

Section 5.7 describes the principles on which laser operation is based, and an outline is given of several ways in which population inversions are obtained in important types of laser. Some lasers operate in CW mode, others in the pulse mode, and a few in both. Table 7.2 summarizes information about some of the lasers discussed.

The large peak powers given in the table for the pulse lasers are one of the important features of these devices, and reflect the short duration of the pulse rather than the total energy available. For example, the 1 MW from the 10-ns pulse dye laser corresponds to only 10 mJ. For a reasonable repetition rate of

Table 7.2 *Properties of some important types of laser (The letters Q or M following a pulse length refer to 'Q-switched' and 'mode-locked' operation, which are discussed in the text)*

Laser	Wavelength	Typical power (peak)	CW or pulse	Excitation method
He–Ne	633 nm	100 mW	CW	Electrical
N_2	337 nm	1 MW	10 ns	Electrical
Ar-ion	458–514 nm	15 W	CW	Electrical
CO_2	10.6 μm	200 W	CW	Electrical
		20 MW	50 ns +1–3 μs tail	
Excimer e.g. KrF	248 nm	20 MW	15 ns	Electrical
Ruby	694 nm	1 mW	CW	Optical
		500 kW	1 ms	
		150 MW	20 ns (Q)	
Nd–YAG	1.06 μm	20 W	CW	Optical
		10 MW	10 ns (Q)	
Nd-glass	1.06 μm	200 MW	50 ns (Q)	Optical
Dye	200–1100 nm	1 W	CW	Optical (laser)
		1 MW	10 ns	
		1000 MW	5 ps (M)	

50 Hz, the *average* power is thus less than 1 W. Nevertheless, peak intensities (i.e. photons per unit area per unit time) are very high indeed.

Two special techniques are available to increase the power and to reduce the length of the pulse in laser sources. The first is *Q-switching*. Normally, stimulated emission starts when the active medium has achieved a large enough population inversion for oscillation to be maintained. Since emission rapidly lowers the inversion below the necessary threshold, the laser output appears as a train of pulses representing swings of the population above and below the threshold. In a Q-switched laser, the laser cavity is prevented from allowing radiation density to build up (e.g. by moving one of the reflectors, or inserting an absorbing medium) until the population inversion has reached a peak level far greater than it would ordinarily attain. When the cavity performance ('Q') is subsequently restored, all the energy accumulated in the inversion can become released as a 'giant' pulse of very short duration, lasting a few nanoseconds. Pulses of even shorter duration can be obtained by *mode locking*. The laser cavity is resonant for radiation that produces standing waves within it. The spectral width of the atomic or molecular transition of the laser process is generally much greater than the bandwidth of the cavity resonance. Thus, a series of longitudinal and transverse 'modes' can be set up in the cavity, each corresponding to an integral number of wavelengths in the optical path. In a free-running multimode laser, the time behaviour of the laser output is irregular, since the different modes have essentially random phases and amplitudes. Forcing the laser modes to oscillate with similar amplitudes and/or with their phases locked yields a train of pulses whose individual components have durations on the time-scale of picoseconds. Pulses lasting a few *femtoseconds* (10^{-15} s) can be obtained by refinements in the technique.

One important feature of laser radiation is its highly monochromatic nature, resulting from the many traversals of a resonant cavity. In the case of a mode-locked laser, the bandwith may be determined ultimately by Uncertainty broadening (cf. p. 39) resulting from the finite length of the pulse. The highest monochromaticity (of the order of 1 part in 10^{12}) is usually achieved with CW lasers. In some laser media, several transitions may be possible, as in the argon ion laser, or the transition involved might even produce a broad-band fluorescence, as in the dye laser. Wavelength selection can be achieved in such cases by replacing the totally reflecting mirror of the laser by a prism or grating so that rays of unwanted wavelengths do not make multiple passes through the laser medium. For a dye laser, this modification means that a selected narrow band of radiation can be obtained, and the device is a *tuneable dye laser*.

Because of the high powers of laser radiation, the possibilities arise of using *non-linear* effects, akin to the multiphoton processes discussed in Section 3.9.

Included in these techniques are *frequency doubling* of a single laser output, and *frequency mixing* of two lasers in certain crystalline dielectric materials. Non-linear processes in gases allow the generation of coherent radiation to be extended into the vacuum ultraviolet region (to wavelengths as short as 100 nm).

The special properties of laser radiation of importance to photochemists are:

(i) High monochromaticity. This property allows highly selective excitation of fine-structure spectral components of a transition.
(ii) Possibility of very short pulses. This property allows very high time resolution in photochemical experiments.
(iii) High peak power resulting from short pulse duration.
(iv) High peak beam density ('intensity') resulting from small beam area.

These last two properties provide the possibility of performing experiments that demand high intensity or 'fluence', such as those involving forbidden transitions, multiphoton processes, absorption saturation phenomena, non-linear effects, etc.

(v) High beam directionality. One use of high directionality is in multipass absorption experiments in which a beam can be made to pass through an absorbing sample many hundreds of times.
(vi) Spatial coherence of beam. Spatially coherent beams are used in subtle spectroscopic experiments that can probe details of electronic states involved in photochemical change.

7.2.4 *Synchrotron radiation*

The vacuum ultraviolet region is an important one, at least for gas-phase photochemistry, because many highly energetic processes can occur, including those involving higher excited states and photoionization. Light sources for this region are generally poor, particularly since continuum sources whose spectral distribution approaches anything like that of a black body at accessible temperatures has negligible output. Low-pressure discharge sources exist, but they produce specific, generally atomic, spectral lines. An exception is provided by *synchrotron radiation*. Electrons are accelerated in circular orbits to a velocity approaching that of light. The laws of electromagnetism require that radiation be emitted, and the resulting synchrotron emission is a structureless continuum spanning the X-ray region to the infrared. The beam is well collimated, and can thus be dispersed with a monochromator to give a narrow band output of high intensity. The radiation is also produced in bursts, with pulse lengths as short as 100 ps, so that time-resolved experiments are possible.

7.3   Classical techniques

7.3.1   *Photolysis and photochemical reaction*

In this section, we look at the nature of experiments designed to follow the overall course, rate, and quantum yield of photolysis or other photochemical change.

A beam of radiation emerging from the combination of lamp and wave-length-selecting device is allowed to fall on to the photochemically active reaction mixture. For quantitative investigations, this reaction mixture is usually contained within a cell which has two parallel plane windows normal to the direction of the incident beam. If the beam is itself near-parallel, the light is absorbed evenly over the whole sample. Any light not absorbed emerges from the rear cell window; in a typical experimental arrangement this transmitted radiation may be allowed to fall on some detector used to measure the intensity (see below). Figure 7.1 shows one common optical arrangement for photochemical experiments in the near-ultraviolet: note that the components are arranged so that the light beam is almost parallel (perhaps slightly convergent) and so that the beam nearly, but not quite, fills the face of the reaction cell.

We come now to the question of measurement of absorbed light intensity. In principle, we need to know the intensity of the light incident on the front

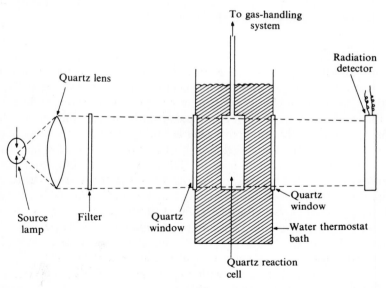

Fig. 7.1   One form of apparatus for 'classical' photochemical experiments using near-ultraviolet radiation.

surface of the absorbing substance and also the fraction of light absorbed. The fraction of light absorbed may, in fact, be calculated directly from the measured concentration and a known extinction coefficient for the absorber, by use of the Beer–Lambert law (Section 2.4, p. 21). Alternatively, the detector used for intensity measurements (cf. Fig. 7.1, and next paragraph) can be used to determine *relative* transmitted intensities in the absence of absorber and with the absorbing substance at the desired concentration in the cell: the fractional absorption at the wavelength of the experiment can be calculated directly.

Light is a form of energy and it may be degraded to heat; if light falls on a blackened surface, the temperature of the surface will rise. The temperature rise is commonly measured by a *thermopile*: this is an assembly of thermo-couples, connected in series, whose front junctions are blackened (Fig. 7.2(a)).

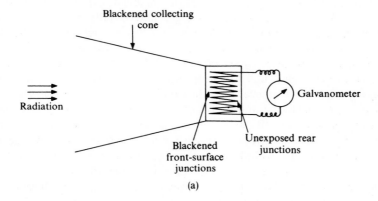

(a)

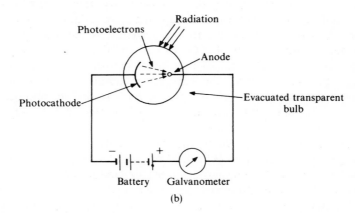

(b)

Fig. 7.2 Radiation detectors: (a) thermopile; (b) photocell.

A temperature difference between the illuminated front junctions and the unexposed rear junctions produces an e.m.f., whose magnitude can be measured. The thermopile may be calibrated against a black body of known temperature, since the overall energy output of such a source is well known. Alternatively, a small heating element may be incorporated in the thermopile block; electrical power is supplied to the heater in order to produce an output e.m.f. identical to that produced by the light, and its magnitude gives a direct measure of the light power. In either case, the energy of the radiation in the photochemical experiment may be converted to units of quanta s^{-1} by use of the Planck relation ($E = h\nu$). Secondary standards, such as standard carbon filament lamps, whose absolute outputs have been determined in separate experiments, are often more convenient to use than direct primary calibration of the thermopile. It is necessary to determine what fraction of radiation from the black body, or from the photolysing beam, is actually collected by the thermopile; the lens system of the photochemical experiment is often arranged so that the light-beam converges and is fully accepted by the thermopile. Corrections must also be made for losses, such as the reflective losses at the rear windows of the cell and the thermostat tank.

Thermopiles are notoriously sensitive to small fluctuations in room temperature and to draughts, and it is more usual to employ a *photocell* in the chemical experiments. A photocell is shown diagramatically in Fig. 7.2(b); it consists of a photocathode and a collector enclosed in an evacuated bulb. Illumination of suitable cathode materials causes the ejection of electrons, and if the collector is charged positively (i.e. if it is the anode) with respect to the cathode, a current will flow in the external circuit. The operating conditions can be chosen so that this current is proportional to the light intensity reaching the photocathode. However, the quantum efficiency of photoelectron emission from the cathode may not be known, and is, in any case, dependent on wavelength. It is necessary, therefore, to calibrate the photocell against a thermopile, or against a secondary standard. The great advantages of the photocell are, first, that it is more sensitive than the thermopile, and, secondly, that the cathode need not respond significantly to long-wavelength radiation, so that small temperature fluctuations are no longer tiresome. Indeed, the cathode material for ultraviolet intensity measurements can be chosen (e.g. pure sodium) so that the photocell does not detect visible light, and stray illumination from the laboratory lighting need not then be rigorously excluded.

An alternative approach to the determination of light intensities is to measure the rate of a photochemical reaction *for which the quantum yield is accurately known*. Chemical systems of this kind are referred to as *chemical actinometers*. The quantum yield for the actinometer itself has, of course, to be determined through absolute (e.g. thermopile) measurements of light intensity. Chemical actinometers are chosen for their insensitivity to wavelength and to the experimental parameters. One of the most convenient materials for use in solution is $K_3Fe(C_2O_4)_3$, a compound known universally in this context as 'potassium ferrioxalate'. Irradiation of ferrioxalate in acid solution results in reduction of Fe^{3+} to Fe^{2+}, and simultaneous oxidation of the $C_2O_4^{2-}$ ion (cf. Section 3.8). In the usual procedure, Fe^{2+} produced is estimated absorptiometrically after formation of a red complex with *o*-phenanthroline. Since no Fe^{2+} iron is initially present, and since the absorption of the red complex is intense, it is easy to demonstrate the formation of, say, $10^{-8}\,mol\,dm^{-3}$ of Fe^{2+}. The quantum yield for Fe^{2+} formation is substantially constant over the wavelength range 254–579 nm, and is virtually insensitive to temperature, solution composition, and light intensity.

One widely used gas-phase actinometer is propanone. For the wavelength region 250–300 nm, and with temperatures above 125 °C and pressures below 50 mmHg, the quantum yield for CO formation is unity. The liquid-phase ferrioxalate actinometer is, however, more generally useful for measurement of light intensities.

7.3.2 *Emission studies*

The quantitative study of luminescence requires the use of certain special techniques, some of which are described in this section. Intensities of fluorescence, phosphorescence, and chemiluminescence are generally much smaller than those of the irradiating light used for photolysis or excitation, and sensitive methods of detection are needed. Photographic records of emission spectra can yield intensity data averaged over the exposure period, as well as providing information about the spectral distribution of the emission. Photoelectric methods are, however, normally used in quantitative investigations because of their superior sensitivity and speed of response. Photocells, such as those described in the last section, can be made to detect radiation of wavelengths as long as about 1300 nm by choice of a suitable cathode (Ag–O–Cs); the short-wavelength limit is set more by the transmission of the cell windows than by the response of the photocathode, and it may be most convenient to coat the front of the cell with a fluorescent material that converts ultraviolet radiation to visible emission detectable with a glass-envelope photocell. Small photocell currents may be amplified by conventional electronic techniques, and low-intensity emission can be detected in this way. Some noise is inevitably introduced by such amplification, and weak emission is best detected with *photomultipliers*. A photomultiplier is effectively a photocell possessing internal amplification that is almost noise-free. Figure 7.3 shows diagramatically the construction of a photomultiplier and the external circuitry employed. A photoelectron ejected from the cathode is accelerated by the electric field towards the first *dynode*; the kinetic energy of impact of the photoelectron on the dynode is sufficient to eject

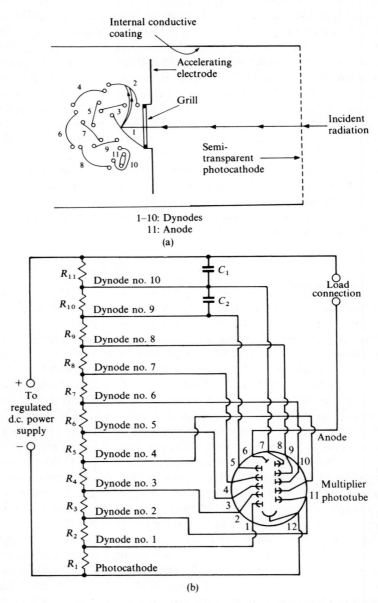

Fig. 7.3 The photomultiplier: (a) construction of one type of photomultiplier; (b) external electrical circuit.

many secondary electrons from the surface material. Each of these secondary electrons is accelerated to the second dynode, each ejects secondary electrons from this dynode, and so on. Current (i.e. numbers of electrons per second) gains of 10^6–10^7 are easily obtained with modern photomultipliers, and since, in principle, only electrons liberated photoelectrically from the cathode can initiate the liberation of secondary electrons at the dynodes, no noise is introduced by the current-amplifying dynode chain. There is, of course, a small number of electrons liberated thermally from the cathode material (and, to an even lesser extent, from the dynodes); cooling the photomultiplier (e.g. in liquid nitrogen) reduces the small thermal 'dark current', and cooled photomultipliers are used for the detection of radiation of exceedingly low intensity.

The spectroscopic nature of luminescent emission may be investigated by the use of a dispersing instrument (e.g. monochromator) together with a photomultiplier to detect radiation. The spectroscopic response of the photo-multiplier must be known in order to determine the true emission spectrum: the spectral sensitivity of photomultipliers is discussed later in connection with absolute calibrations. Fluorescence or phosphorescence *excitation spec-tra* are obtained by monitoring the emission intensity (preferably within a narrow wavelength band) as the wavelength of the exciting light is altered: the true excitation spectrum is obtained only if the intensity of exciting light is constant at all wavelengths, and if the intensity is not constant, appropriate corrections must be applied to the observed excitation spectrum.

Fluorescence and phosphorescence are almost always observed at right angles to the direction of the irradiating beam. Scattered light of the incident radiation can be troublesome, especially if the luminescence is weak, and the presence of dust or polycrystalline solids often precludes the study of lumi-nescence lying at wavelengths near that of the exciting light. Reflections of the exciting light by the cell faces can be virtually eliminated by suitable cell design: Fig. 7.4 shows the use of 'Rayleigh Horns' in a fluorescence cell. If an irradiated system exhibits both fluorescence and phosphorescence, it may be difficult to establish the contribution to emission made by each process, on the basis of the steady-state spectral characteristics alone. The basic techni-que used to differentiate between the two phenomena was devised by Bequerel in 1859. Early forms of the apparatus employed mechanical devices such as rotating drums to alternately expose the sample to exciting radiation or to allow the experimenter to observe it, and thus to see if any luminescence persisted after illumination was cut off. The method is a simple time-resolved one (with a time resolution, depending on drum rotation speed, of about 1 ms), and the present-day technique also uses time-resolved intensity analy-sis, of the kind that will be presented in Section 7.5.

Absolute quantum efficiencies for fluorescence or phosphorescence can be calculated from measurements in the *same arbitrary units* of absorbed and

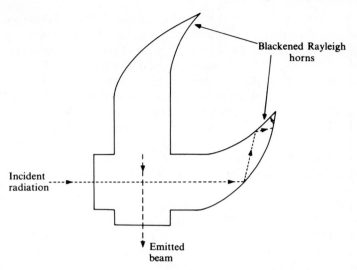

Incident
radiation

Blackened Rayleigh
horns

Emitted
beam

Fig. 7.4   Cell for fluorescence studies. The illustration shows multiple reflections of
the incident beam into the blackened Rayleigh horn.

emitted intensities. Allowances must be made for the differences in spatial and spectral distributions of exciting and emitted radiation, and the spectral response curve of the detector must be known. The directed exciting beam may be scattered for comparison with isotropic emitted radiation either by a matt surface, or, better, by a protein solution of calculable scattering power. A simplification of the corrections required for the spectral distribution of emitted light may be achieved by allowing the emitted radiation and the scattered exciting beam to fall successively on a suitable fluorescent substance that converts all incident radiation *to its own fluorescence spectrum with uniform quantum efficiency.* Such a device is called a 'quantum counter', and it is clearly necessary that the fluorescence quantum yield be wavelength-independent for all wavelengths in the emitted and exciting radiation; it is *not,* however, necessary to know the absolute value of the quantum yield for the counter material. One substance frequently employed is a solution of rhodamine-B. The use of the quantum counter removes problems concerned both with the spectral sensitivity of the detector and with the spectral distributions of exciting and emitted light, since the detector always receives the same fluorescence spectrum of the counter material regardless of the excitation wavelength. It may be noted that the relative spectral sensitivity curve of a detector can be determined by comparison of response, at a series of wavelengths, to a monochromatic scattered beam viewed directly, and to the fluorescent emission excited by the radiation in a quantum counter.

Direct evaluation of quantum yields of emission processes by measurement of *absolute* emission (and absorption) intensities is possible, although the low intensity of many emission processes makes such measurements difficult. The absolute intensities may be determined either by the primary thermopile standard, or by a previously calibrated photomultiplier. The potassium ferrioxalate chemical actinometer may also be employed for absolute emission intensity measurements because of its high sensitivity.

The absolute quantum yield of a luminescent process may, of course, be assessed by comparison of emitted intensities from the sample and from a substance whose emission quantum yield is already accurately known. Although this method begs the question of how the quantum yield was originally determined for the standard substance, it may in practice be the most convenient and rapid technique. Several standards have been suggested. One useful substance is the sodium salt of 1-naphthylamine-4-sulphonic acid; dilute, oxygen-free, solutions in glycol exhibit a quantum yield (determined by one of the 'absolute' methods) close to unity. A similar technique is finding increasing application in measurements of absolute intensities in gas-phase chemiluminescence. The spectral distribution and absolute emission efficiency of the 'air-afterglow' chemiluminescence (reaction (4.35), p. 96)

$$O + NO + M \rightarrow NO_2^* + M \qquad (7.1)$$

is known from experiments using the potassium ferrioxalate actinometer. It is a simple matter to measure the intensity of the air afterglow resulting from the reaction of known concentrations of O and NO and to compare it with the intensity, at any wavelength, of a process whose efficiency is to be measured. Since the geometrical conditions are identical for 'known' and 'unknown' chemiluminescence, it is unnecessary to estimate the fraction of the total emitted light that is received by the detector.

7.4 Identification and estimation of photochemical intermediates

The development over the past few decades of techniques to detect and identify the reactive intermediates involved in photochemical change, and to measure their concentrations, has been one of the major factors in our increased understanding of the details of photochemistry. The kinds of species involved are: the atoms, radicals, and ions that are the primary fragments of photolysis; the excited states of these fragments; and the excited states, including triplets, of the absorber that are first produced and that participate in fluorescence, phosphorescence, and radiationless transitions (IC and ISC). It is the capability of examining these reactive species on a short time-scale that has opened up the possibilities of the sophisticated time-resolved experiments that are the subject of the next section. Conversely, it is often the time-resolved experiments themselves that permit the interrogation of a photochemical system at a time soon enough after the absorption event that the reactive species of interest is still present. In this section, we review briefly some of the more important techniques that are available for the investigation of intermediates in order to provide an introduction to the discussion of time-resolved studies. This is not the place to discuss the theoretical background to the spectroscopy used, but rather to show which methods can profitably be employed. One theme that appears many times is how lasers have made older kinds of spectroscopic experiment easier, as well as how they have made possible entirely new sorts of study.

Optical spectroscopy, especially in the visible and ultraviolet regions, often provides both the most sensitive way of detecting an intermediate and the most unequivocal method of identifying it. Particularly in the case of a species with resolvable fine structure in the spectrum, the spectroscopic detail may enable description of the chemical nature, the structure, and the electronic state of the species. *Absorption spectroscopy* can be used in conjunction with a white-light source and a spectrograph to provide a photographically-recorded survey spectrum of the absorbing species in a reaction system. Alternatively, a monochromator with photoelectric detection can be used to scan over the spectral region. Many species of interest show quite strong absorptions, the requirement being that there is another higher-lying state to which an electric dipole transition is allowed. In this way, for example, triplet species may be observed by their triplet–triplet absorptions. In general, the individual absorption features are stronger, the sharper the spectral features (because, for example, in a broad feature, the same absorbed fraction of the incident energy is spread over a wide band). Atoms have particularly strong allowed absorption lines because of this effect. For quantitative work, it is usual to choose a wavelength for measuring the absorbance where the absorption is strong and free from interference from other species. Lasers may be used as the light source in absorption experiments and *laser absorption* studies using tuneable dye lasers are very successful, especially with species possessing sharp absorption features (such as atoms and small radicals), because the laser radiation is itself highly monochromatic and has a narrow bandwidth. Increased absorption may be obtained by making the light-beam traverse the sample many times using suitably arranged mirrors in a *multipass absorption* experiment, and lasers again lend themselves admirably to this purpose as a result of the small divergence of the laser beam that can be achieved. In some cases, it may be possible to construct a light source that is spectrally matched to the absorption feature being studied. For example, electric discharge lamps containing the appropriate gases can be made to emit the resonance spectral lines (i.e. to the ground state, in principle from the first excited state) of many atoms and simple free radicals. These lines obviously exactly match the wavelengths of the absorptions of the ground-state species if they are present in the reaction mixture, and *resonance absorption* will result. So long as the spectral linewidths of source and absorber match, the sensitivity of absorption can be high, and the specificity is also excellent because it is unlikely that another, chemically unrelated, species would accidentally have a comparably intense absorption at the exact wavelength emitted by the lamp.

Optical emission can be used to study excited species so long as an optical transition can occur with sufficient intensity to a lower state. Thus, the excited species of fluorescence or chemiluminescence are well suited to this kind of study, while the weaker emission of phosphorescence restricts the usefulness of emission in investigating the triplets of organic molecules, especially in time-resolved experiments.

Emission studies have some advantages over absorption studies. In particular, absorption by an intermediate present at low concentration implies a small change in the relatively large intensity of the probing beam. 'Noise' resulting from adventitious fluctuations in the light intensity, as well as the statistical nature of the beam photons, limits the sensitivity attainable in an absorption experiment. In an emission experiment, on the other hand, there is no radiation produced unless the excited species is present. Statistical limitations still restrict the precision with which concentrations can be measured, but the ultimate practical sensitivity of the emission experiment is usually much higher than that of the absorption one. For this reason, emission is frequently harnessed to study species initially in their ground state by deliberately pumping them optically to a higher, emitting, state. Where appropriate, the resonance discharge lamps described earlier can be used to excite *resonance fluorescence* in atoms (e.g. H, O, Cl) and radicals (e.g. OH). The fluorescence is isotropic, and so may be detected off-axis from the exciting beam. Tuneable lasers can be used with great success as pumps in this application. Lasers offer much greater versatility than the discharge lamps, particularly in terms of the range of molecular species that can be pumped (e.g. OH, NO_3, CH_3O, C_2H_5O), and the high power available in the exciting radiation ensures high sensitivity. *Laser Induced Fluorescence* (LIF) has become a most valuable tool in the study of gas-phase intermediates, and sensitivities of the order of 10^6 molecule cm^{-3} ($< 2 \times 10^{-15}$ mol dm^{-3}) are obtained.

Lasers can also be put to use as pumps in Raman studies. Laser–Raman spectroscopy has found several applications in studying reaction intermediates in photochemical processes. The high intensity and monochromaticity of the laser radiation lend the Raman method a sensitivity that it does not have with conventional light sources, and make it suitable for time-resolved investigations of intermediates. *Resonance Raman Spectroscopy* (RRS) becomes possible with tuneable light sources. When the wavelength of the Raman exciting radiation matches a strong absorption band in the species under study, the Raman scattering is up to six orders of magnitude more intense than in the ordinary, non-resonant, case. One specially valuable variant of laser–Raman spectroscopy is *Coherent Anti-Stokes Raman Spectroscopy* (CARS), which depends on non-linear behaviour in the presence of intense radiation, and involves the mixing of several waves. High sensitivity is obtained because the observations are performed on an emission rather than on an absorption system. Unwanted scattering of the exciting light limits the sensitivity of ordinary Raman studies, but in the CARS experiment, none of the pumping radiation lies close in wavelength to the CARS emission, so scattered radiation from the pump lasers can be filtered out.

There are some remarks about the comparative performance of absorption, emission, and Raman studies that are worth making. Although emission studies are generally more sensitive than absorption, they are limited to species that have an excited state that lives long enough for spontaneous emission, with an A-factor of not more than ca. $10^8 \, s^{-1}$, to compete effectively with predissociation or other radiationless loss processes over which the experimenter has no control (but see Section 7.6). Furthermore, the lifetime of the emission process itself imposes a limit on the minimum useable time-scale of time-resolved experiments (say to around 10^{-8} s). Absorption and Raman interactions with electromagnetic radiation occur over approximately a single cycle of the wave, or around 10^{-15} s in the ultraviolet region. Reactive intermediates can thus be probed on the femtosecond time-scale, which exceeds the limit of almost all current time-resolved experiments.

Photoionization methods have been used for some time in the identification of intermediates, but the use of lasers in ionization studies has greatly enhanced the range of applicability. The underlying idea is that, with mono-energetic photons of the correct energy, it should be possible to ionize an intermediate (e.g. CH_3) without fragmenting and ionizing its precursor (e.g. CH_4), or it should be possible to ionize an excited state of a species without ionizing lower-lying states. The sensitivity is high, because ions are only produced when the intermediate is present, and mass filters may be employed to provide mass identification of the ions. Multiphoton ionization by lasers (see MPI and REMPI, pp. 60–61) permits the ionization of a wide variety of species without the necessity of using vacuum ultraviolet sources, and very large ionization efficiencies can be achieved at high laser intensities.

Magnetic resonance methods find a considerable range of application in the study of photochemical intermediates. *Electron Spin Resonance* (ESR) itself has been used to study not only doublet radicals, but triplet species as well. In the gas phase, species with orbital momentum (e.g. $O_2\,^1\Delta_g$) also produce paramagnetic resonance, but the principle area of application has been in condensed-phase studies. One drawback to direct ESR studies is the limited speed of response attainable (around 1 μs), which is largely imposed by the characteristics of the microwave cavity. *ESR Spin Echo* methods have been used to reach a time-scale of around 50 ns, but the fastest of the ESR techniques is *Optically Detected Magnetic Resonance* (ODMR), which is capable of time resolution of the order of a few nanoseconds. The method is one of many *double-resonance* techniques. The microwave transition is recognized not by the absorption measured directly in the microwave region, but by some effect, such as change in absorption or fluorescence in the visible region, that is a consequence of changed interactions as the different spin states are populated. We have referred already (pp. 55–56) to the CIDNP and CIDEP techniques for studying the behaviour of radical pairs; CIDNP employs polarization in the recombined product molecules, while CIDEP looks at anomalous polarizations in the radicals themselves to infer the kinetic behaviour of the radical species. Optically detected magnetic resonance experiments complement these studies of radical pairs, and provide information about dissociation, and the events immediately following, in condensed-phase systems. *Laser Magnetic Resonance* (LMR), a technique for gas-phase studies, has some elements in common with the other magnetic resonance experiments described so far, in that it relies on the Zeeman splitting of levels in a magnetic field. Here, however, transitions within the Zeeman levels are not studied, but rather the variation of Zeeman splitting with magnetic field is used to 'tune' an infrared-active absorption in an atom or radical into resonance with a fixed-frequency laser. Laser magnetic resonance has been used to study species such as OH, ClO, HO_2, CH_3O, and CH_2OH.

7.5 Time-resolved photochemistry

The archetypal method of performing time-resolved experiments in photochemistry is *flash photolysis*. This technique was developed initially by Norrish and Porter in the 1950s in an attempt to identify the reactive intermediates present in photochemical systems. The stationary concentrations of atoms, radicals, or excited species present in a static system are normally too low for the intermediates to be detected by their absorption spectra. However, if an extremely-high-intensity flash source is used, then the transient concentrations of intermediates may be sufficiently large for spectroscopic observation. Further, the changes with respect to time of the intermediate concentration may be followed by means of the absorption spectrum, and kinetic data, such as radical lifetimes, may be obtained. This use of time-resolved spectroscopy is often known as *kinetic spectroscopy*. (Kinetic spectroscopy may also be used to monitor continuously the concentrations of suitable reactants and final products as a function of time after the flash.) The mechanisms of many photochemical reactions have finally been elucidated with the help of the information given by flash photolysis experiments about the nature and reactivity of intermediates.

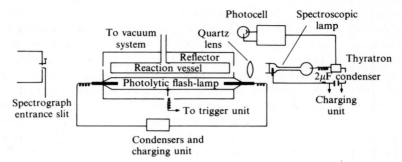

Fig. 7.5   Schematic diagram of flash photolysis apparatus. (From Wayne, R. P. (1969). In *Comprehensive chemical kinetics* (eds C. H. Bamford and C. F. H. Tipper) Vol. I, Fig. 3, p. 284. Elsevier, Amsterdam.)

Figure 7.5 shows an early form of the apparatus. A rare-gas flash-lamp, of the type described in Section 7.2, is placed parallel to a quartz cell containing the reaction mixture. Both lamp and cell are surrounded by a reflecting enclosure. The lamp is connected to a charged bank of capacitors, and the discharge triggered. A small amount of light from the photolysis flash is allowed to fall on to a photocell, which is connected to an electronic delay unit. The delay unit then triggers a second, low-power, 'spectroscopic' flash-lamp, the light from which is directed down the length of the reaction cell and ultimately on to the slit of a spectrograph. A photographic record is thus made of the absorption spectrum of the reaction vessel contents at a fixed time (determined by the electronic delay) after the photolysis flash. The reaction vessel is then refilled with fresh reactant and the experiment is repeated with a different delay time. In this way a series of spectra, corresponding to the different delay times, is obtained. The experiment is a 'pump-and-probe' procedure.

Flash spectroscopy in the form just described is most convenient for survey investigation of the formation and decay of reactive intermediates, although it may be used for quantitative kinetic work by microdensitometry of the photographic plate. A more convenient modification for kinetic experiments, *flash spectrophotometry*, employs a broad-band CW source (e.g. a tungsten–halogen lamp) and monochromator combination set at the wavelength of absorption of the species of interest. The transmitted intensity is monitored as a function of time with a fast photomultiplier, and the output is either displayed on an oscilloscope whose sweep is triggered by the photolytic flash, the trace being photographed, or, more usually in modern practice, stored directly by an electronic *transient recorder*. Figure 7.6 shows a photographic record of absorbance as a function of time in such an experiment.

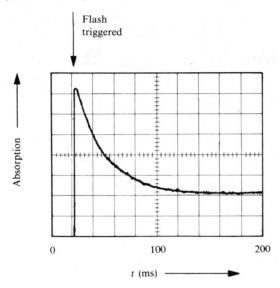

Fig. 7.6 Transient absorption of the triplet state of anthracene in a polymer glass. The figure shows a photograph of an oscilloscope trace from a flash photolysis–kinetic spectrometry experiment. (Diagram kindly provided by Hi-Tech Scientific Ltd, Salisbury.)

Laser sources may, of course, replace the lamp–monochromator combination.

Luminescent lifetimes may be studied in a manner wholly analogous to absorption flash spectrophotometry, the monitoring light source being omitted. A further obvious variant is the use of resonance fluorescence (RF) or laser induced fluorescence (LIF) to probe the concentrations of non-emitting intermediates produced by the photolytic flash.

A high-power rare-gas photolytic lamp produces a flash whose duration is not less than about 1 μs. Deconvolution of the pumping and absorption (or emission) time–intensity profiles permits some improvement of the time-resolution of the experiments, but to reach the true nanosecond time-scale, some modifications are needed. For studying fast luminescence (e.g. fluorescence), it may be possible to use a modified discharge lamp filled with gases such as air or nitrogen that give a much sharper pulse if the electrical circuitry is properly designed. These lamps do, however, give very low output intensities. It is therefore necessary to carry out the experiment repeatedly (tens of thousands to millions of times) to obtain data of the required precision. The photomultiplier output is taken to a storage or sampling oscilloscope or, more often, to a digital electronic storage device, or *multiscaler*, referred to in this context as a *Computer of Average Transients* (CAT). The repetition rate of the experiment can itself exceed 10 000 times per second, so that the repetitive nature of the experiment may not impose too great a penalty. An alternative way of recording the data in luminescence experiments is to use *single-photon correlation*, which relies on there very rarely being more than one emitted photon produced for each flash of the exciting discharge. An electronic timer is started by the exciting flash, and stopped when a fluorescent photon is received by the detecting photomultiplier; a record of the time taken is sent to a computer. As the experiment is repeated, the computer builds up a record of the number of fluorescent photons that arrive in each time interval. With a large enough number of repeated experiments, the record is thus of the fluorescence intensity as a function of time, which is the information required for kinetic analysis.

Nanosecond *photolysis* experiments generally rely on pulsed lasers as the primary source, because the fast discharge lamps are too weak. In flash spectroscopy, the probe light source also needs to be fast. One method that is sometimes adopted uses a fluorescent material pumped by a second laser triggered after an appropriate delay. The fluorescence can be sufficiently broad-banded for spectroscopy (which the laser is not), and its intensity–time profile is determined by its fluorescence lifetime. For nanosecond flash spectrophotometry, a suitable light source may be a conventional flash discharge, with a flash duration of hundreds of microseconds, so that the source intensity is high during the period when absorption is being recorded.

Experiments on the picosecond time-scale and shorter begin to require rather different approaches. The photolytic or exciting flash has to be obtained from a mode-locked laser, with a single pulse selected from the pulse train. Although picosecond flash spectroscopy relies on the double-flash pump-and-probe method, the monitoring probe beam is usually derived from the photolytic flash, and the short time delay that is necessary is introduced by using the finite velocity of light. The monitoring beam is sent on an adjustable, longer, optical path. It may be spectrally broadened for absorption experiments (e.g. CCl_4 converts a small fraction of $\lambda = 1060$ nm from a Nd-glass laser to broad-band radiation), or be frequency converted for other diagnostic techniques such as CARS. Several specialized methods also exist for studying emission on the picosecond time-scale. One of these relies on an electronic modification of the *streak camera*. In its mechanical implementation, a streak camera employs a fast-moving photographic film to record intensity of a focused beam or spot as a function of time. In the electronic version, the image is first allowed to fall on the photocathode of a special kind of imaging photomultiplier related to a television camera tube. The photo-electrons produced are then swept rapidly by applying a ramp voltage to plates inside the tube. A luminescent phosphor screen with relatively long persistence, and which can be photographed, can be used to record the arrival positions of the swept electron image, or the electron arrival can be captured electronically for subsequent analysis. The method is that of *electron optical chronoscopy*. An alternative method for studying picosecond-resolved fluorescence uses a shutter based on the Kerr effect (rotation of the plane of polarization of light in the presence of an electric field) induced optically by a gating laser pulse. Yet another method (*fluorescence correlation spectroscopy*) mixes an optically-delayed portion of the exciting laser pulse with the fluorescent emission in a non-linear crystal (cf. p. 159) to produce shorter-wavelength radiation, which is then detected with a fast photomultiplier. The intensity of this radiation depends on the intensities of both fluorescence and probe beams, so that the variation of intensity with delay of the probe beam is simply related to the variation of fluorescence intensity with time.

Experiments on the millisecond and microsecond time-scale provide information about the rates of bimolecular reaction of photolytic fragments and excited states, as well as about the emission of radiation from triplets in phosphorescence. Nanosecond experiments are able to probe the fluorescent emission from singlets, as well as intersystem crossing, while picosecond studies yield kinetic data about geminate recombination, energy exchange, vibrational relaxation, and some of the slower internal conversions and isomerizations. Investigations on the femtosecond time-scale are beginning to be reported, although it should be borne in mind that in one femtosecond light travels only 300 nm, or one wavelength! Experiments on these time-scales probe the act of absorption of radiation and the very first stages of utilization of the energy in promoting chemical and physical change.

7.6 High-resolution photochemistry

Much of this book has been about electronically excited species populated above their thermal equilibrium populations, directly by the absorption of light or in subsequent processes. Considerable interest attaches to the nascent distribution of energy amongst electronic, vibrational, and rotational internal modes, and in translational motion, immediately following a photochemical event such as photodissociation. 'Nascent' in this context means that no significant redistribution of energy has taken place subsequent to the event, so that the disposition of energy amongst the various modes can be used to infer the dynamics of the event. In a chemical reaction, the molecular system passes smoothly from reactants to products through an intermediate species; the field of reaction dynamics is concerned with how the laws of physical dynamics determine the approach of reactants and the departure of the products. In one sense, reaction dynamics is at the heart of all chemical transformations, and a better understanding of the dynamics represents a better understanding of the transformation itself. *State-to-state kinetics* provides some of the experimental information for theoretical approaches to reaction dynamics by attempting to investigate the rates of processes involving reactants in specific internal quantum states, and with specific velocities and starting coordinates, to form products in equally well-defined quantum states, velocities, and angular distributions and momenta. Of course, most chemistry is not performed with state-selected species, but rather with something approaching thermally equilibrated statistical distributions. Nevertheless, these statistical distributions are made up of the individual states, so that the assembly of state-to-state processes makes up the whole reaction. Techniques are just now emerging that permit the 'high-resolution' study of state-to-state photochemistry, and it is appropriate to conclude this chapter on experimental methods with a brief mention of some of the developments that will fuel theoretical physical photochemistry in the next few decades.

Photodissociation dynamics is a particularly promising part of reaction dynamics to study, because the interaction between a photon and a molecule is a 'half-collision', in which the initial properties of the photon are as well defined as possible. Theoretical models of the dissociation process can be tested against the results of experiments that measure the nascent energy and momentum disposal. The models frequently indicate that there is strong sensitivity to the quantum states in the absorbing molecule, so that the fullest experimental tests will require state-selected reactants as well as state identification in the products. The observable quantities that are most sensitive to the dynamics of dissociation seem to be the rotational energy disposal, and the angular distributions and orientations of the fragments, and many of the more sophisticated studies have attempted to look at these parameters.

Preparing reactants in selected quantum states may need the application of *molecular beam* techniques, which do not directly concern us here. Alternatively, the products of certain thermal or photochemical reactions may be populated preferentially in particular states. With polyatomic species, rotational, and even vibrational, levels in the reactants may be so 'crowded' together with the populations present at ambient temperatures that selection or excitation of individual levels is impossible. One very successful technique that has been adopted in such cases has been to use *supersonic expansion* of the sample in a beam or jet. Expansion leads to a conversion of the random translational energy of the molecules to the forward-directed motion, and energy transfer can reduce the effective temperature of the molecules to a few kelvin, so that only the very lowest levels of vibrational and rotational states remain populated. In this way, the absorption spectrum becomes much simpler, and individual levels can be pumped optically.

Investigation of the nascent energy disposal requires at least that intermolecular collisions do not redistribute the energy amongst the modes. Very low gas pressures may be needed, which is why the kind of work being discussed is essentially confined to gas-phase systems. Collision-free molecular beams may offer an even better way of avoiding collisions. One important technique that employs beams is *photofragment time-of-flight spectroscopy*. The time taken for photodissociation fragments to reach a detector placed remotely at the end of a drift tube allows the translational velocity, and hence energy, of the fragments to be deduced. The difference between photon energy and molecular dissociation energy then reveals the disposition of fragment energy between translational and internal modes. Sometimes several peaks appear for a particular fragment, usually representative of the different vibrational levels formed, and sometimes also indicating that more than one electronically excited state has been created. The width of individual peaks provides a measure of the rotational distribution in the fragments. In a development of the technique, the detector assembly can be moved in relation to the incoming molecular beam in order to yield the important angular information about the fragmentation process. Detailed data about the fragment rotational distributions can usually be obtained only by fast spectroscopy. Absorption, photoionization, and CARS, as described in Section 7.4, have all been applied in this context.

While intermolecular energy exchange can be minimized by avoiding collisions, intramolecular energy exchange within the same molecule cannot. In these circumstances, the only way to obtain the nascent energy distributions is to probe the dissociating system at times short enough after the photon has been absorbed for energy exchange to be impossible. It is in meeting this aim, and in understanding the exchange processes themselves, that the very fast time-resolved experiments described in the last section are so valuable.

The experimental study of the nuclear motions of photodissociation offers a real challenge, because molecular structural changes occur over internuclear distances of a few tenths of a nanometer, corresponding to times on a femtosecond time-scale. One interesting approach to this problem has been to use the spectroscopic effects of the nuclear motions as an indicator of the time-dependent behaviour. In effect, the requirement for time-resolution has been transferred to the amplitude–frequency spectral measurement. As a specific example, consider the molecule O_3. On absorption of a photon, it predissociates in about one vibration. It certainly is not normally considered to be a fluorescent molecule (cf. pp. 39, 76). However, a very small fraction of the molecules does emit (about 1 in 10^6), and with intense laser illumination and sensitive detection an emission spectrum *can* be recorded. The first interesting feature of the fluorescence is that it contains unusually long vibrational progressions. As the molecule comes apart, it sweeps through all molecular displacements so that transitions to all bound levels have high Franck–Condon probabilities (cf. pp. 31–32). More important from the point of view of dissociation dynamics is that intensities of the lines observed are representative of the molecular motions of the *excited* state, and thus of the dissociation process. Dissociation of O_3 by ultraviolet radiation affords a clear example of how a qualitative understanding of the dynamics can be arrived at simply. No bands of the bending vibrational mode are seen in the emission spectrum, suggesting straight away that the bending mode is not involved in the early stages of the reaction. Furthermore, only transitions involving *even* levels of the antisymmetric stretching vibration are seen. This result is interpreted in terms of the symmetry of the dissociation process.

The experiments described in the last few sections are aimed at providing a description of photophysical and photochemical interactions at the most detailed and fundamental level. For many chemists, the important aspects of photochemistry are what it does in nature, what new chemicals it can make, and in what processes it can be harnessed, and these are the concerns of our next chapter.

Bibliography

Rabek, J. F. (1982). *Experimental methods in photochemistry and photophysics*, Two volumes. John Wiley, Chichester and New York.

Calvert, J. G. and Pitts, J. N., Jr (1966). *Photochemistry*, Chapter 7. John Wiley, Chichester and New York.

Parker, C. A. (1969). *Luminescence of solutions*, Chapter 3. Elsevier, Amsterdam.

Finlayson-Pitts, B. J. and Pitts, J. N., Jr (1986). *Atmospheric chemistry, Part 3, Experimental kinetic, mechanistic and spectroscopic techniques.* John Wiley, Chichester and New York.

Okabe, H. (1978). *Photochemistry of small molecules*, Chapter 3. John Wiley, Chichester and New York.

Special topics
Section 7.2

Phillips, R. (1983). *Sources and applications of ultraviolet radiation.* Academic Press, New York.

Andrews, D. L. (1986). *Lasers in chemistry.* Springer-Verlag, Berlin.

West, M. A. (ed.) (1977). *Laser in chemistry.* Elsevier, Amsterdam.

Hochstrasser, R. M. and Johnson, C. K. (1985). Lasers in biology. *Laser Focus* **21**, 100.

Winick, H. (1987). Synchrotron radiation. *Sci. Am.* **257**(5), 72.

Kunz, C. (1979). *Synchrotron radiation: techniques and applications.* Springer-Verlag, Berlin.

Dutuit, O., Tabche-Fouhaile, A., Nenner, I., Frohlich, H., and Guyon, P. M. (1985). Photodissociation processes of water vapor below and above the ionization potential. *J. Chem. Phys.* **83**, 584. [This paper describes the use of synchrotron radiation in a photochemical study.]

Sections 7.4 and 7.5

Wayne, R. P. (1969). The detection and estimation of intermediates. In *Comprehensive chemical kinetics* (eds C. H. Bamford and C. F. H. Tipper) Vol. 2. Elsevier, Amsterdam.

Norrish, R. G. W. and Thrush, B. A. (1956). Flash photolysis and kinetic spectroscopy. *Q. Rev. Chem. Soc.* **10**, 149.

Kevan, L. and Schwartz, R. N. (eds) (1979). *Time domain electron spin resonance.* John Wiley, Chichester and New York.

Weissman, S. I. (1982). Recent developments in electron paramagnetic resonance: transient methods. *Ann. Rev. Phys. Chem.* **33**, 301.

Symons, M. C. R. and McLauchlan, K. A. (1984). From trapped radicals to transients. *Faraday Discuss. Chem. Soc.* **78**, 7.

Closs, G. L., Miller, R. J., and Redwine, O. R. (1985). Time-resolved CIDNP: applications to radical and biradical chemistry. *Acc. Chem. Res.* **18**, 196.

El-Sayed, M. A. (1974). Phosphorescence–microwave multiple resonance spectroscopy. *Adv. Photochem.* **9**, 311.

Molin, Y. N., Anisimov, O. A., Melekhov, V. I., and Smirnov, S. N. (1984). Optically detected electron spin resonance studies of electrons and holes involved in geminate recombination in non-polar solutions. *Faraday Discuss. Chem. Soc.* **78**, 289.

Davies, P. B. (1981). Laser magnetic resonance spectroscopy. *J. Phys. Chem.* **85**, 2599.

Atkinson, G. H. (1982). Time-resolved Raman spectroscopy. *Adv. Infrared Raman Spectrosc.* **9**, 1.

Terner, J. and El-Sayed, M. A. (1985). Time-resolved resonance Raman spectroscopy of photobiological and photochemical transients. *Acc. Chem. Res.* **18**, 331.

Fleming, G. R. (1986). *Chemical applications of ultrafast spectroscopy.* Oxford University Press, Oxford.

Hilinski, E. F. and Rentzepis, P. M. (1983). Chemical applications of picosecond spectroscopy. *Acc. Chem. Res.* **16**, 224.

Simon, J. D. and Peters, K. S. (1984). Picosecond studies of organic photoreactions. *Acc. Chem. Res.* **17**, 277.

Hopkins, J. B. and Rentzepis, P. M. (1986). Advances in picosecond spectroscopy. *Chem. Br.* **22**, 236.

Fleming, G. R. (1986). Subpicosecond spectroscopy. *Ann. Rev. Phys. Chem.* **37**, 81.

Peters, K. (1987). Picosecond organic photochemistry. *Ann. Rev. Phys. Chem.* **38**, 253.

Grischkowsky, D. R. (ed.) (1985). Femtosecond optical interactions. *J. Opt. Soc. Am.* **B2**, 582.

Section 7.6

Leone, S. R. (1984). State-resolved molecular reaction dynamics. *Ann. Rev. Phys. Chem.* **35**, 109.

Simons, J. P. (1984). Photodissociation: a critical survey. *J. Phys. Chem.* **88**, 1287.

Reisler, H. and Wittig, C. (1986). Photoinitiated unimolecular reactions. *Ann. Rev. Phys. Chem.* **37**, 307.

Shapiro, M. and Bersohn, R. (1982). Theories of the dynamics of photodissociation. *Ann. Rev. Phys. Chem.* **33**, 409.

Leone, S. R. (1982). Photofragment dynamics. *Adv. Chem. Phys.* **50**, 255.

Smalley, R. E. (1983). Dynamics of electronically excited states. *Ann. Rev. Phys. Chem.* **34**, 129.

Jackson, W. M. and Okabe, H. (1986). Photodissociation dynamics of small molecules. *Adv. Photochem.* **13**, 1.

Greene, C. H. and Zare, R. N. (1982). Photofragment alignment and orientation. *Ann. Rev. Phys. Chem.* **33**, 119.

Levy, D. H. (1980). Laser spectroscopy of cold gas-phase molecules. *Ann. Rev. Phys. Chem.* **31**, 197.

Zewail, A. H. (1983). Picosecond-jet spectroscopy and photochemistry. *Faraday Discuss. Chem. Soc.* **75**, 315.

Leone, S. R. (1983). Infrared fluorescence: a versatile probe of state-selected chemical dynamics. *Acc. Chem. Res.* **16**, 88.

Bernstein, R. B. (1982). *Chemical dynamics via molecular beam and laser techniques.* Oxford University Press, Oxford.

Lee, Y. T. and Shen Y. R. (1980). Studies with crossed laser and molecular beams. *Phys. Today* **33**, 52.

Lee, E. K. C. (1977). Laser photochemistry of selected vibronic and rotational states. *Acc. Chem. Res.* **10**, 319.

Imre, D., Kinsey, J. L., Sinha, A., and Krenos, J. (1984). Chemical dynamics studied by emission spectroscopy of dissociating molecules. *J. Phys. Chem.* **88**, 3956.

Foth, H. J., Polanyi, J. C., and Telle H. H. (1982). Emission from molecules and reaction intermediates in the process of falling apart. *J. Phys. Chem.* **86**, 5027.

8

Photochemistry in action

8.1 Applied photochemistry

Earlier chapters of this book have been concerned with the fundamental processes of photochemistry. In this last chapter, we look at some of the many ways in which photochemistry has an impact on our lives. Natural photochemical phenomena have contributed to the evolution of life as we know it, and permit its continued existence on Earth. Applications of photochemistry, ranging from photography to photomedicine, are of great importance to us. Only a limited number of examples can be presented here, but those selected should illustrate the diversity of the applications.

8.2 Atmospheric photochemistry

8.2.1 *Origin and evolution of the atmosphere*

Photochemical reactions have performed a determining role in the evolution of the atmosphere and of life on Earth. Our understanding of primary photochemical processes permits reasonable speculation about the history of the atmosphere; the 'investigation' of the Earth's palaeoatmosphere (fossil atmosphere) has in turn suggested solutions to several 'puzzles' concerning the Earth's geology. The forms and ecology of life that were viable at any time in the past were directly dependent on the constitution of the atmosphere at that period; we shall see that, conversely, processes involving living organisms exert a major influence on atmospheric composition. It is this interrelation between atmospheric and biological evolution that makes the study of the Earth's palaeoatmosphere, and comparison with the present-day atmospheres of other planets, particularly rewarding. In this section we put forward one view of the development of the Earth's atmosphere.

Much evidence shows that Earth was without a primordial atmosphere. For example, the abundances of the rare gases in the contemporary atmosphere lie between 10^{-10} and 10^{-6} of their cosmic abundances. It has been shown that the quantities of gases liberated as a result of volcanic activity, and from slow decay of solid radioactive elements, are sufficient to account for our atmosphere. However, *oxygen is not released from volcanic effluents*, and the primitive atmosphere must have contained N_2, CO_2, and H_2O as its most important constituents, together with traces of reducing gases such as H_2 and CO.

In the absence of life, the main source of O_2 must have been photolysis of water by short-wavelength ultraviolet radiation:

$$H_2O + h\nu \rightarrow 2H + O \qquad (8.1)$$

$$O + O(+M) \rightarrow O_2(+M) \qquad (8.2)$$

The production of O_2 by this mechanism is limited by the 'shadowing' effect of the O_2 formed. Molecular oxygen is likely to have been distributed in the atmosphere above the H_2O vapour, and to have absorbed the radiation responsible for photodissociation of water (say at $\lambda < 195$ nm). This argument can be extended quantitatively to show that the upper limit for oxygen concentration in the primitive atmosphere was less (probably much less) than 10^{-3} of the present atmospheric level (PAL). Absorption of ultraviolet radiation by CO_2 would add to the shadowing and reduce further the limiting O_2 concentration. Geological evidence is consistent with $[O_2] < 10^{-3}$ PAL: the incomplete oxidation of early sedimentary materials suggests sedimentation in a reducing atmosphere, and such oxides as were formed could have resulted from oxidation by relatively small amounts of *ozone* formed close to the Earth's surface by the three-body process:

$$O + O_2 + M \rightarrow O_3 + M \qquad (8.3)$$

(Present-day ozone distributions are discussed later in this section.)

How could our atmosphere come to have more than 20% of oxygen in it, while our nearest neighbours, Venus and Mars, have less than 0.1%? Earth possesses an atmosphere that for hundreds of millions of years appears to have been disregarding the laws of physics and chemistry. Minor oxidizable constituents of our atmosphere, such as methane, ammonia, hydrogen, carbon monoxide, and nitrous oxide, survive in the presence of large concentrations of oxygen. Thermodynamic considerations would suggest virtually complete oxidation of these components. Earth's peculiar behaviour is a consequence of the existence of life on it. Biological processes, acting together with physical and chemical change, determine the composition of our atmosphere. Conversely, our unique atmosphere seems essential for the support of life in many of the forms that we know it. Oxygen, the unexpected gas of our atmosphere, is almost entirely the result of biological activity. Not only does biology provide the atmospheric oxidant, but it also continually provides the oxidizable minor gas 'fuels'. Biological processes bring about the thermodynamic disequilibrium of our atmosphere, and, in effect, reduce its entropy.

Energy is needed for this entropy reduction, and virtually all of it is supplied by radiation from the Sun, mediated by the biota.

Photosynthesis is the only known process that can have caused the rise of $[O_2]$ from 10^{-3} PAL to 1 PAL. Photosynthesis is the subject of Section 8.3, and for the time being we need only note that the process involves consumption of carbon dioxide and water, and the concomitant *liberation of oxygen*. In the present atmosphere, all O_2 passes through the photosynthetic process in a few thousand years, a period extremely short by geological standards, and photosynthesis is clearly an efficient source of O_2. The build-up of oxygen in the atmosphere is dependent on a rate of O_2 production (mainly a result of photosynthesis) in excess of the rate of loss (resulting from oxidation, respiration, etc.). In the first stages of progress from $[O_2] \sim 10^{-3}$ PAL, photosynthetic activity (at about present densities) must have covered a few per cent of the (present) continental areas before there was a net positive balance contributing to atmospheric oxygen.

Pre-biological oxygen concentrations in the atmosphere are important in two ways. Organic molecules are susceptible to thermal oxidation and photo-oxidation, and are unlikely to have accumulated in large concentrations in an oxidizing atmosphere. The low pre-biological oxygen concentrations thus seem essential to the development of the organic precursors of life. Living organisms can develop mechanisms that protect against oxidative degradation, but they are still photochemically sensitive to short-wavelength ultraviolet radiation. Macromolecules, such as the proteins and nucleic acids that are characteristic of living cells, are damaged by ultraviolet radiation of wavelength shorter than about 290 nm. In our atmosphere, the oxygen itself is able to filter out solar ultraviolet radiation with wavelengths shorter than about 230 nm. For wavelengths between 230 and 290 nm, however, some other protection must be afforded. Luckily, there is a suitable absorbing species in our atmosphere, so that organisms can live on dry land more or less exposed to the filtered Sun's rays. The species is ozone, O_3, derived photochemically from O_2 in the atmosphere (see p. 186). The amount of ozone in the atmosphere, and its altitude distribution, will depend on the concentration of the precursor oxygen, and will thus have altered markedly as the atmosphere evolved. Concentrations of ozone are also controlled by the rates of loss processes for the molecule. Loss is regulated by catalytic cycles involving other trace gases of the atmosphere, such as the oxides of nitrogen, that are themselves at least partly of biological origin (see p. 189). We have already noted that the oxygen in the Earth's atmosphere comes largely from biological sources. Now we see that the ozone, needed as a filter to protect life, has its concentration determined not only by the biologically-generated oxygen needed for its production, but also by the biologically-generated trace gases that play a part in its destruction. Such observations have led Lovelock to the idea of *Gaia* (Earth Mother), in which climate and the chemical composition of the Earth's surface and atmosphere are kept at an optimum by and for the biosphere. It is at least clear that our study of the evolution of Earth's atmosphere must also include a cursory look at the evolution of life.

Life on Earth appears to extend *at least* as far back as the earliest sedimentary rocks, fossils of microscopic organisms suggesting the presence of abundant life 3.5 Gy BP (3.5×10^9 y Before Present). Bacterial cells lacking nuclei initially contributed oxygen to the atmosphere. Familiar animal, plant, and fungal cells have nuclei, but require oxygen in relatively large quantities. A revolution occurred as oxygen became more plentiful in the atmosphere, as nucleated cells, and then animal and plant life, emerged. Respiration and large-scale photosynthesis thus become of importance at this stage, probably when $[O_2] \simeq 10^{-2}$ PAL, and some time between 2.0 and 0.57 Gy BP, thus encompassing the dawn of the Cambrian period (0.57 Gy BP). Following the opening of the Cambrian, the complexity of life is known to have multiplied rapidly, and the foundations for all modern phyla were laid. 'Advanced', non-microscopic, lifeforms were found ashore by the Silurian age (420 My BP), and by the early Devonian, only 30 My later, great forests had appeared. Soon afterwards, amphibian vertebrates ventured onto dry land.

According to one interpretation, due first to Berkner and Marshall, the evolution of O_2, and hence of protective O_3, controlled the migration of life onto land. At low atmospheric O_2, *liquid* water, at a depth of say 10 m, will filter out much of the damaging ultraviolet radiation, while allowing photo-synthetically active visible light to reach living organisms. Life in the oceans seems improbable at this stage, since organisms would be brought too near the surface by mechanical motions, and it would probably be confined to the safety of stagnant pools and lakes. When O_2 and O_3 had built up yet further, the ultraviolet zone of lethality would be restricted to a thin layer at the ocean surface, so that life could spread to entire ocean areas, thus greatly enhancing photosynthetic activity. As the oxygen content of the atmosphere moved towards its present level, enough O_3 was available for no liquid water to be needed for protection, and life could finally be supported on dry land, probably at $[O_2] > 10^{-1}$ PAL.

A major question surrounding the interpretation of Berkner and Marshall is whether the biological evolutionary events were causally related to the atmospheric changes that undoubtedly occurred. If they were, then some kind of feedback mechanism of the kind postulated for Gaia (see p. 183) may have been in operation, since the atmospheric evolution was certainly me-diated by the biota. The kind of problem encountered in relating biological and atmospheric evolution is exemplified by the development of shells by metazoan organisms. Because the shells are relatively impervious to oxygen, shelled organisms require dissolved oxygen that would be in equilibrium with $> 10^{-1}$ PAL in the atmosphere, so that the critical level of O_3 for biological protection would have been passed when the organisms appeared abundantly in the Cambrian period (570 My BP). But life was probably not firmly established on dry land for a further 170 My (Late Silurian). Thus, the possibility exists that the ozone screen may already have been established before the Silurian, and that it was not directly linked with the spread of life onto land. Resolution of this question will require further information. In the first place, it is necessary to know the time history of growth of O_2 and O_3 in the atmosphere. Laboratory studies of photochemical kinetics make possible the calculation of the O_3 levels that accompanied those O_2 levels suggested for the early atmosphere by the geological evidence. Figure 8.1 summarizes the results of such a calculation, the evolutionary events that are part of the evidence being marked along the oxygen concentration curve.

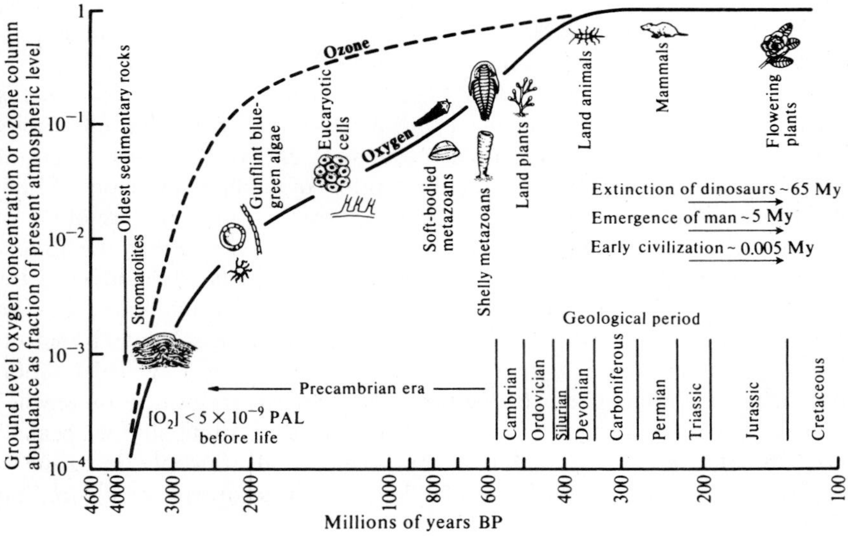

Fig. 8.1 Evolution of oxygen, ozone, and life on Earth. In the absence of life, surface oxygen concentrations are unlikely to have exceeded $\sim 5 \times 10^{-9}$ of the present value. The build-up of oxygen to its present level is largely a result of photosynthesis. Early organisms would have found high oxygen concentrations toxic, but eucaryotic (nucleated) cells require at least several per cent of the present level for their respiration. Soft-bodied metazoans could have survived at similar oxygen levels, but the reduced surface oxygen uptake area available once the species had developed shells must mean that the concentration was approaching one-tenth of its current value about 570 My ago. Considerations such as these are used in drawing up the oxygen growth curve. Ozone concentrations can be derived from a photochemical model. Life could not have become established on land until there was enough ozone to afford protection from solar ultraviolet radiation. (From Wayne, R. P. (1985). *Chemistry of atmospheres*. Oxford University Press, Oxford.)

8.2.2   *The stratosphere*

Simple thermodynamic arguments suggest that temperature should decrease with altitude in the atmosphere. In the Earth's atmosphere the temperature drops by about 6.5 K for every kilometre of height increase, for roughly the first 15–20 km above the surface, but above this height the temperature begins to increase again. The reversal in the temperature trend constitutes a *temperature inversion*: it results mainly from solar photodissociation of ozone, and the subsequent exothermic chemical reactions that we shall discuss shortly. The lowest region of the atmosphere has colder air lying on top of warmer air so that convection can lead to rapid vertical mixing: the region is named the *troposphere* after the Greek for 'turning'. In the second region, the warmer air lying on top of colder air results in great vertical stability, and the region is named the *stratosphere* after the Latin for 'layered'. The troposphere and the stratosphere are divided by the *tropopause*. Ozone in the contemporary stratosphere is the next topic of our survey.

Our survival depends, no less than our evolution did, on our being protected by atmospheric ozone from short-wavelength solar ultraviolet radiation. Furthermore, the ultimate source of energy for many reactions occurring in the atmosphere is the absorption of sunlight by ozone. Much interest is attached, therefore, to the measurement and interpretation of present-day concentrations and altitude distributions of ozone in the atmosphere. Rocket, balloon, and satellite experiments make possible direct measurements of the ozone altitude profile, and these determinations can be compared with the predictions of hypothetical reaction schemes based on laboratory kinetic data. Figure 8.2 shows the results of a typical balloon investigation of atmospheric ozone concentration. The concentration reaches a maximum at an altitude of around 27 km which is quite sharply peaked (note that the abscissa is a logarithmic scale), and atmospheric ozone is frequently described as consisting of a 'layer' in the stratosphere, centred on 25–30 km.

The basic processes that establish the ozone layer were described by Chapman as long ago as 1930. The important 'oxygen-only' reactions are

$$O_2 + h\nu \rightarrow O + O \quad \text{for } \lambda < 242.4 \text{ nm} \tag{8.4}$$

$$O_3 + h\nu \rightarrow O_2 + O \quad \text{for } \lambda < 1180 \text{ nm} \tag{8.5}$$

$$O + O_2 + M \rightarrow O_3 + M \tag{8.3}$$

$$O + O_3 \rightarrow O_2 + O_2 \tag{8.6}$$

Because reactions (8.5) and (8.3) can interconvert atomic oxygen and ozone, O and O_3 are identified as the family of 'odd oxygen'. Reaction (8.4) creates two odd oxygens, and reaction (8.6) destroys two, while reactions (8.5) and (8.3) themselves obviously leave the odd-oxygen concentration unaltered,

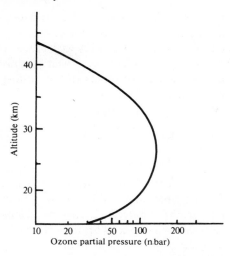

Fig. 8.2 Variation of atmospheric ozone concentration with altitude, as measured with instruments carried aloft by a balloon. (From Hudson, R. (ed.) (1981). *The stratosphere 1981*. World Meteorological Organization, Geneva.)

although they do affect the ratio of [O] to [O_3]. After sunset, atomic oxygen concentrations fall rapidly at altitudes below about 40 km, since the source reactions (8.4) and (8.5) are cut off, but the sink processes (8.3) and (8.6) remain. Ozone is thus neither formed nor destroyed at night, and diurnal variations in [O_3] are small at these altitudes. Higher in the atmosphere, diurnal changes are pronounced as daytime photolysis of O_3 becomes faster, and conversion of O back to O_3 becomes slower because of the lower pressure, the rate of reaction (8.3) being proportional to the *square* of the pressure.

The simple four-reaction scheme predicts the layer structure found in the atmosphere. At high altitudes, there is much short-wavelength ultraviolet radiation capable of dissociating molecular oxygen, but relatively little of the O_2 itself. Low in the atmosphere, there is plenty of O_2, but short-wavelength radiation is absent, because it has already been filtered out by the O_2 and the O_3 lying above. Solar ultraviolet energy absorbed by the O_2 and O_3 is ultimately liberated as heat, in part through the exothermic chemical reactions (8.3) and (8.6). It is this heating that gives rise to the stratospheric temperature inversion.

Proper calculation of the ozone profile, using rate parameters for the reactions determined in the laboratory, shows that the predicted profile has the same general shape as the measured one. However, the calculated absolute concentrations are all higher, by a factor of up to four or five, than the true atmospheric ones. The problem arises because the loss process, reaction (8.6), has an activation energy of 18.4 kJ mol^{-1}, and is too slow at stratospheric temperatures (say 220–270 K) to balance the production of ozone at the correct concentration. It is now well established that reaction (8.6) can be catalysed by trace constituents in the atmosphere. The idea is summed up in the reaction schemes

$$X + O_3 \rightarrow XO + O_2 \tag{8.7}$$

$$XO + O \rightarrow X + O_2 \tag{8.8}$$

$$\text{Net:} \quad \overline{O + O_3 \rightarrow O_2 + O_2}$$

The reactive species X is regenerated in the second step, so that its abundance is not affected by its participation in odd-oxygen removal. Several species have been suggested for the catalytic 'X' in the atmosphere. The most important for the natural stratosphere are $X = H$ and OH, $X = NO$, and $X = Cl$; the catalytic cycles are then said to involve HO_x, NO_x, and Cl_x. Typical pairs of reactions are

$$H + O_3 \rightarrow OH + O_2 \tag{8.9}$$
$$OH + O \rightarrow H + O_2 \tag{8.10}$$

$$OH + O_3 \rightarrow HO_2 + O_2 \tag{8.11}$$
$$HO_2 + O \rightarrow OH + O_2 \tag{8.12}$$

$$NO + O_3 \rightarrow NO_2 + O_2 \tag{8.13}$$
$$NO_2 + O \rightarrow NO + O_2 \tag{8.14}$$

$$Cl + O_3 \rightarrow ClO + O_2 \tag{8.15}$$
$$ClO + O \rightarrow Cl + O_2 \tag{8.16}$$

and further catalytic cycles that destroy odd oxygen have also been identified. All the cycles have activation energies for the individual steps that are lower than the activation energy of the direct $O + O_3$ reaction. Whether or not the catalytic reactions are actually *faster* than the direct reaction at stratospheric temperatures will depend on the relative concentrations of XO and O_3. Throughout much of the stratosphere, loss of odd oxygen turns out to be dominated by the NO_x cycle for altitudes up to about 45 km. The ClO_x cycle also becomes faster than the direct reaction at altitudes greater than about 30 km, and above 50 km, the HO_x cycles become the most important loss mechanisms.

All three catalytic families, HO_x, NO_x, and Cl_x appear to be present in the 'natural' atmosphere unpolluted by Man's activities. Precursors of the catalytic species have sources at the Earth's surface (supplemented in the case of NO_x by direct conversion of the N_2 and O_2 of the atmosphere at high altitudes). These precursors have to be transported through the troposphere to the stratosphere. Amongst the species of importance are H_2O, CH_4, N_2O, and CH_3Cl, which are converted to the catalyst radicals in the stratosphere. Photolysis of ozone by *ultraviolet* radiation leads to the formation of electronically excited fragments

$$O_3 + h\nu_{UV} \rightarrow O(^1D) + O_2(^1\Delta_g) \qquad (8.17)$$

and the excited oxygen atom can react with H_2O, CH_4, or N_2O in reactions that are thermodynamically unfavourable for ground-state atomic oxygen (cf. Chapter 6, p. 130)

$$O(^1D) + H_2O \rightarrow OH + OH \qquad (8.18)$$

$$O(^1D) + CH_4 \rightarrow OH + CH_3 \qquad (8.19)$$

$$O(^1D) + N_2O \rightarrow NO + NO \qquad (8.20)$$

and CH_3Cl can be photolysed directly

$$CH_3Cl + h\nu \rightarrow Cl + CH_3 \qquad (8.21)$$

to provide an entry to the Cl_x cycle (another entry involves reaction with OH).

The stratosphere is very dry, probably because water from the troposphere has to pass through the 'cold trap' at the tropopause, and CH_4 constitutes more than a third of the total $[H_2O] + [CH_4]$ in the lower stratosphere. Reaction (8.19) is therefore an important source of OH, especially since the oxidation of the CH_3 radical (to CO) also yields two or three more odd-hydrogen species. Both N_2O and CH_4 are the result of biological activity (mostly microbial) on the Earth's surface. The main contribution to CH_3Cl is again biological, this time in the oceans, although burning of vegetation and some volcanic eruptions are additional sources.

Numerical 'models' of the atmosphere that incorporate the chemistry just described predict ozone concentration–altitude profiles well, especially if they allow for horizontal and vertical transport of the chemical species. An additional test of the theory can be provided by measuring concentrations of the reactive intermediates that are the catalytic chain carriers. Figure 8.3 shows an example of such a test. Atomic oxygen concentrations were measured in the same balloon flight as $[O_3]$, and the figure compares the experimental values of $[O]/[O_3]$ with those predicted by a numerical model. The agreement is remarkably good, especially in view of the difficulties of obtaining the experimental data.

Increased understanding of the role played by trace gases in determining atmospheric ozone concentrations has also led to an awareness that Man might inadvertently alter ozone concentrations by releasing catalytically active materials. Pollutants introduced into the stratosphere would have a

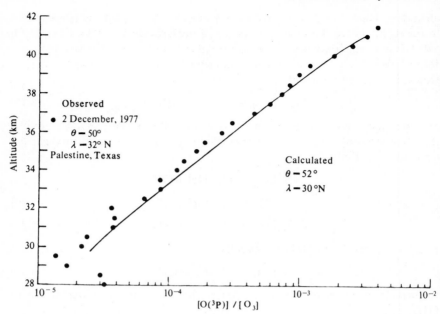

Fig. 8.3   Ratio of concentrations of atomic oxygen to ozone. (●) Experimental points; (—) calculated values. (Source as for Fig. 8.2.)

lifetime for physical removal by transport of several years because of the vertical stability that results from the temperature inversion. They might therefore build up to globally-damaging levels, reducing stratospheric ozone with biological consequences at ground level, such as increased incidence of skin cancer. Initial concern, in the early 1970s, centred on supersonic strato-spheric transport (SST) aircraft, such as Concorde. Such aircraft could inject NO_x, produced from N_2 and O_2 in the high temperatures of the jet engines, directly into the stratosphere. Current numerical models indicate that ozone reductions due to SSTs are negligible, partly because the present fleet of aircraft is so small and partly because the aircraft fly low in the stratosphere where the NO_x cycle has relatively little effect on $[O_3]$. Another source of increased stratospheric NO_x would be increased N_2O production in the biosphere resulting from intensive use of fertilizers. While perturbations due to the use of SSTs may be regarded as discretionary, agricultural use of fertilizers may be essential if populations continue to grow. Doubling the N_2O concentration is predicted to give a global ozone depletion of between 9 and 16%, although such large increases in N_2O are improbable in the near future. A more immediate problem seems to be the release of fluorinated chlorocarbons (CFCs), such as dichlorodifluoromethane, CF_2Cl_2 (CFC-12), and trichlorofluoromethane, $CFCl_3$ (CFC-11). The CFCs are extremely inert chemically, and are valuable as aerosol propellants, as refrigerants, as blowing agents for plastic foam production, and as solvents. The uses of the CFCs all lead ultimately to atmospheric release, and it appears that the quantities of CFCs in the troposphere are equal, within experimental error, to the total amount ever manufactured. Tropospheric inertness is thus confirmed, and lifetimes of up to hundreds of years are indicated. Only one escape route is available to the CFCs, and that is upward transport to the stratosphere. Sufficiently short-wavelength ultraviolet radiation penetrates to the stratosphere to photolyse the CFCs, the process involving liberation of atomic chlorine, as exemplified in reaction (8.22) for CFC-12

$$CF_2Cl_2 + hv \rightarrow CF_2Cl + Cl \qquad (8.22)$$

The Cl atoms thus contribute to ozone destruction in the ClO_x cycle, reactions (8.15) and (8.16), and chlorine of man-made origin now dominates over the natural CH_3Cl contribution. Many models predict significant ozone depletions from the release of CFCs. A feature of most predictions is that the depletions will get larger over one or more decades even if CFC release is curtailed now, and that the full recovery of ozone concentrations may take up to 100 years. There is already (1987) some evidence that global ozone levels are decreasing somewhat faster than expected. In recent years a new phenomenon (the 'ozone hole') has been observed in ozone levels in the Antarctic, where extremely large depletions occur each year in October. The extent of the depletion seems to have been growing each year, and there is considerable speculation that the 'hole' may be associated with the increasing atmospheric load of CFCs.

8.2.3 *The troposphere*

About 90% of the total atmospheric mass resides in the troposphere, and the bulk of the minor trace gas burden is found there also. The Earth's surface acts as the main source of the trace gases, although some NO_x and CO may be produced in thunderstorms. Hydroxyl radicals dominate the chemistry of the troposphere in the same way that oxygen atoms and ozone dominate stratospheric chemistry. Free-radical chain reactions initiated by OH oxidize H_2, CH_4 and other hydrocarbons, and CO, to CO_2 and H_2O. The reactions thus constitute a low-temperature combustion system. The free-radical chain processes are photochemically driven, although stratospheric ozone limits the solar radiation at the Earth's surface to wavelengths longer than 280 nm. At these wavelengths, the most important photochemically active species are O_3, NO_2, and HCHO. All three can yield OH (or HO_2) indirectly, and thus initiate the oxidation chains. Ozone photolysis is, however, a critical step, since the other photolytic processes owe either their origin or their importance to it. Although only 10% of the total atmospheric ozone is found in the troposphere, all *primary* initiation of oxidation chains in the natural atmosphere depends on that ozone. Some ozone is transported to the troposphere from the stratospheric ozone layer, but a mechanism also exists for generation of ozone in the troposphere itself. If NO_2 is present, then NO_2 photolysis (at $\lambda \leqslant 400$ nm)

$$NO_2 + h\nu \rightarrow O + NO \tag{8.23}$$

is a source of atomic oxygen that can form ozone in reaction (8.3)

$$O + O_2 + M \rightarrow O_3 + M \tag{8.3}$$

The NO can itself be oxidized back to NO_2, as we shall see shortly, so that the formation of O_3 is not stoicheiometrically limited by the supply of NO_2 molecules initially present.

A first understanding of tropospheric photochemistry may be gained by considering methane as the only hydrocarbon present, and taking as our starting point the artificial situation where no CH_4 has yet been oxidized. Hydroxyl radicals must then be derived from ozone photolysis (at $\lambda \leqslant 310$ nm), in the way already described for the stratosphere

$$O_3 + h\nu_{UV} \rightarrow O(^1D) + O_2(^1\Delta_g) \tag{8.17}$$

$$O(^1D) + H_2O \rightarrow OH + OH \tag{8.18}$$

Attack of OH on CH_4 yields methyl radicals, and a sequence of oxidation steps ensues that we will follow, for the time being, to the formation of HCHO (formaldehyde).

$$OH + CH_4 \rightarrow CH_3 + H_2O \tag{8.24}$$

$$CH_3 + O_2 + M \rightarrow CH_3O_2 + M \tag{8.25}$$

$$CH_3O_2 + NO \rightarrow CH_3O + NO_2 \tag{8.26}$$

$$CH_3O + O_2 \rightarrow HCHO + HO_2 \tag{8.27}$$

$$HO_2 + NO \rightarrow OH + NO_2 \tag{8.28}$$

Two very important features are displayed by this scheme. First, reactions (8.26) and (8.28) both provide a route for the oxidation of NO back to NO_2, and thus to a replenishment of tropospheric ozone through reactions (8.23) and (8.3). Secondly, the reactions as written are cyclic, the OH radical chain carrier being regenerated.

The aldehyde product of reaction (8.27) can itself be photolysed in the troposphere; the major photolytic pathway at $\lambda \leqslant 338$ nm yields two radical fragments that enter into further reactions

$$HCHO + h\nu \rightarrow H + HCO \tag{8.29}$$

$$HCO + O_2 \rightarrow CO + HO_2 \qquad (8.30)$$

$$H + O_2 + M \rightarrow HO_2 + M \qquad (8.31)$$

Finally, we may follow the oxidation through yet another step. Carbon monoxide reacts with OH

$$OH + CO \rightarrow H + CO_2 \qquad (8.32)$$

so that the ultimate product is CO_2. The H atom re-enters the oxidation chain via reactions (8.31) and (8.28). In the unpolluted atmosphere, roughly 70% of the OH reacts with CO, and 30% with methane itself.

The oxidation steps that we have written for methane obviously have their analogues for higher hydrocarbons, but in all cases the reactions depend on the switch between peroxy- (RO_2) and oxy- (RO) radicals in an interaction with NO. Oxides of nitrogen are therefore a central part of the oxidation scheme, because they both effect the switch and are the ultimate source of ozone, and thus of OH radicals. Natural sources of NO_x include microbial actions in the soil, which produce NO as well as N_2O. Oxidation of biogenic NH_3, initiated by OH radicals, would be another significant source of NO_x. Lightning discharges appear to be responsible for less than 10% of the total NO_x budget.

Man's activities lead to the release of many kinds of pollutants to the troposphere, and the chemistry of all polluted urban atmospheres necessarily involves some photochemical processes. However, in this section we describe a form of pollution whose origin is essentially photochemical—the photochemical 'smog' found typically in Los Angeles.

There are several reasons why Los Angeles suffers particularly from photochemical smog. As we shall see later, the pollution derives largely from automobile exhaust gases. Los Angeles has the world's greatest traffic density, and sunshine is consistently intense, so that smog formation is favoured. In addition, the meteorological features of the Los Angeles basin, surrounded as it is by a ring of high mountains and the sea, lead to stagnation of the air and trapping of pollutants.

The characteristic pollutants are ozone and nitrogen dioxide, together with a host of organic compounds. Concentrations of O_3 and NO_2 are so high that the ozone can easily be detected by smell, and a heavy load of particles leads to a brown haze in the air. Damage to materials such as rubber, damage to vegetation, reduction in visibility, and increased incidence of respiratory disease are recognized consequences of the pollution; the most immediately obvious effect of photochemical smog is eye irritation caused by substances such as formaldehyde, acrolein, and peroxyacetyl nitrate (PAN).

Figure 8.4 shows the diurnal variations in concentrations of several pollutants in Los Angeles during a smoggy day. Nitric oxide, NO, is initially present, but is oxidized *after dawn* to nitrogen dioxide; ozone appears only

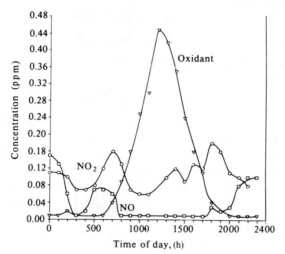

Fig. 8.4 Variations in concentration of oxidant (mainly ozone) and oxides of nitrogen during the course of a smoggy day in Southern California. (From Finlayson-Pitts, B. J. and Pitts, J. N., Jr (1977). *Adv. Environ. Sci. Technol.* **7**, 75.)

after most of the nitric oxide has been oxidized. Small amounts of nitric oxide are known to be liberated, together with hydrocarbons, in automobile exhaust gases, and laboratory studies in test chambers ('smog chambers') of the effect of ultraviolet irradiation of such gases in air show concentration–time dependences (Fig. 8.5) qualitatively similar to those for urban air pollution (Fig. 8.4). The presence of ultraviolet radiation is seen in both cases to be necessary to effect oxidation of NO and of hydrocarbons.

The third-order process

$$2NO + O_2 \rightarrow 2NO_2 \tag{8.33}$$

would be far too slow, at the low NO concentrations involved, to account for much NO_2 formation, and the only reasonable inorganic chemical reaction that can lead to sufficiently rapid oxidation of NO is

$$NO + O_3 \rightarrow NO_2 + O_2 \tag{8.34}$$

The absence of O_3 until almost all the NO has been oxidized (Figs 8.4 and 8.5) is consistent with reaction (8.34) being of major importance, at least for loss of O_3. However, as described above, the only tropospheric source of O_3 starts with the photolysis of NO_2, so that inorganic processes alone cannot account for the oxidation of NO to NO_2. Indeed, the smog chamber experiments show clearly that NO_2 is not formed if the hydrocarbon component is left out of the irradiated mixture.

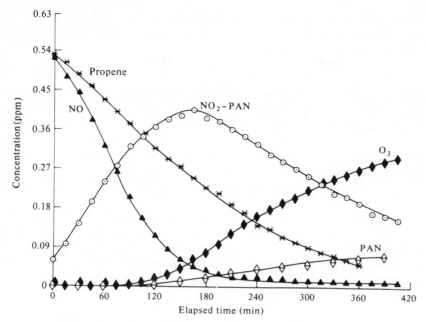

Fig. 8.5 Concentration–time profiles of the major primary and secondary pollutants during irradiation of 0.53 ppm propene and 0.59 ppm NO_x in 1 atm of purified air in an evacuable smog chamber. (Source as for Fig. 8.4.)

The oxidation of both NO and the hydrocarbons is, in fact, a consequence of an exaggerated form of the chemistry already described for the natural troposphere. For an alkane RCH_3, the sequence may therefore be written

$$OH + RCH_3 \rightarrow RCH_2 + H_2O \tag{8.35}$$

$$RCH_2 + O_2 \rightarrow RCH_2O_2 \tag{8.36}$$

$$RCH_2O_2 + NO \rightarrow RCH_2O + NO_2 \tag{8.37}$$

$$RCH_2O + O_2 \rightarrow RCHO + HO_2 \tag{8.38}$$

$$HO_2 + NO \rightarrow OH + NO_2 \tag{8.28}$$

Net: $RCH_3 + 2NO + 2O_2 \rightarrow RCHO + 2NO_2 + H_2O$

(the third body, M, has been dropped from the addition reaction (8.36), as the reactions are second-order at near-atmospheric pressure for radicals larger than CH_3). Although the mechanism is presented in terms of attack of OH on an alkane as fuel, alkenes react with OH even more rapidly than the alkanes.

Writing the net change brought about in the oxidation emphasizes the conversion of NO to NO_2, but still says nothing about the origin of the radicals. The ozone–water photochemical source described for the natural atmosphere (reaction (8.17) followed by reaction (8.18)) may be supplemented by the photolysis of nitrous acid, HONO,

$$HONO + h\nu \rightarrow OH + NO \tag{8.39}$$

and of formaldehyde, reaction (8.29), in polluted atmospheres. Smog chamber experiments support the idea that aldehydes are important. Addition of formaldehyde to hydrocarbon–NO_x–air mixtures causes NO to be converted to NO_2 much more rapidly, and ozone appears earlier and at higher concentrations. Attack of OH on aldehydes leads to the formation of RCO radicals

$$OH + RCHO \rightarrow RCO + H_2O \tag{8.40}$$

which can themselves form R radicals and acids, RCOOH. The degradation from RCH_2 to R radicals is accompanied by aldehydes and acids down to HCHO and HCOOH, as found in photochemical smog. The RCO radicals are of interest in another way, since they can add to O_2 in a reaction analogous to reaction (8.36)

$$RCO + O_2 \rightarrow RCO \cdot O_2 \tag{8.41}$$

to yield peroxyacyl radicals, and then by further addition of NO_2

$$RCO \cdot O_2 + NO_2 \rightarrow RCO \cdot O_2 \cdot NO_2 \tag{8.42}$$

to form peroxyacyl nitrates. Peroxyacetyl nitrate (PAN), $CH_3CO \cdot O_2 \cdot NO_2$, is an important component of photochemical smog, acting as an irritant of the respiratory system and the eyes, and being highly toxic to plants (phyto-toxic).

Aerosols of particulate matter are found in many kinds of air pollution, most obviously, of course, in the contamination associated with burning coal (e.g. in London before the 'Clean Air Act'), but also in photochemical smog. The presence of suspended particles in the air leads to a serious reduction in visibility. Although the origin of the particulate matter in photochemical smog is not clear, it appears to involve the oxidative polymerization of hydrocarbons (possibly aromatic); laboratory studies have shown that aerosols can be formed by the irradiation of automobile exhaust gases. Aerosols are also produced in a form of natural air pollution, found in many parts of the world, but notably in the southwestern USA. A photochemically induced contamination of the atmosphere by particulate matter gives rise to a haze or smokiness over regions possessing high densities of trees such as pines or citrus fruit. Terpenes can be oxidized by ozone to give particulate matter,[†]

[†]The formation of particulate matter in this way can be demonstrated dramatically in the laboratory by squeezing a piece of orange *peel* near a flask of ozonized oxygen—a bluish cloud appears in the flask.

and it seems that the atmospheric aerosols are formed in this way by reactions of terpenes liberated from the trees.

8.3 Photosynthesis

Photosynthesis is perhaps the most important of the many interesting photochemical processes known in biology; not only was the evolution of the Earth's atmosphere dependent on it, but also animal life derives energy from the Sun via photosynthesis by eating plants. It is estimated that the total mass of organic material produced by green plants during the biological history of the Earth represents 1% of the planet's mass, and that photosynthesis fixes annually the equivalent of ten times mankind's energy consumption. In this section we shall discuss photosynthesis with respect to green plants, although it should be noted that there are certain other photosynthetic organisms (e.g. some bacteria) in which the essential photochemistry may be somewhat modified.

From the point of view of organic synthesis, the overall process consists of the formation of carbohydrates by the reduction of carbon dioxide:

$$n\text{CO}_2 + n\text{H}_2\text{O} \xrightarrow{h\nu} (\text{CH}_2\text{O})_n + n\text{O}_2 \qquad (8.43)$$

The essence of the process is the use of photochemical energy to split water and, hence, to reduce CO_2. Molecular oxygen is liberated in the reaction, although it appears at an earlier stage in the sequence of steps than the reduction of CO_2. True photochemical processes appear to produce compounds of high chemical potential, which can 'drive' the synthetic sequence from CO_2 to carbohydrate in a cyclic fashion.

Reaction (8.43) is thermodynamically very improbable in the dark ($\Delta H^{\ominus} = 470$ kJ and $\Delta G^{\ominus} = 500$ kJ per mole of CO_2 converted). Production of one molecule of oxygen requires the transfer of four electrons, and four electrons are also needed to reduce one CO_2 molecule to carbohydrate

$$2\text{H}_2\text{O} \rightarrow \text{O}_2 + 4\text{e} + 4\text{H}^+ \qquad (8.44)$$

$$4\text{e} + 4\text{H}^+ + \text{CO}_2 \rightarrow (\text{CH}_2\text{O}) + \text{H}_2\text{O} \qquad (8.45)$$

Reaction (8.43), with $n = 1$, is the sum of reactions (8.44) and (8.45), and it is evident that if each photon absorbed can lead to the transfer of one electron, then a minimum of four photons are needed for each CO_2 molecule converted. Experimental measurements of the quantum yield indicate that *eight* photons are needed in reality, suggesting that two photons are utilized for each electron transfer, and thus pointing to a two-step process with relatively long-lived intermediates connecting the steps. Other evidence, to be presented shortly, is in accord with this view.

Calvin has elucidated the mechanism of the actual carbon cycle—research for which he received the Nobel Prize. The details of the cycle do not concern us; the important feature for our purpose is the input of energy by (and reducing power of) the specific compounds adenosine triphosphate (ATP) and the reduced form (NADPH) of nicotinamide adenine dinucleotide phosphate (NADP). Figure 8.6 shows the formulae of ATP and NADP. The synthetic carbon cycle can, in fact, be driven by ATP and NADPH in the presence of all initial enzymes and substrates, *but in the absence of light*. Thus the primary and secondary photochemical acts appear to result ultimately in the formation of ATP and NADPH by the photophosphorylation of adenosine diphosphate (ADP) and the reduction of NADP. We may represent these

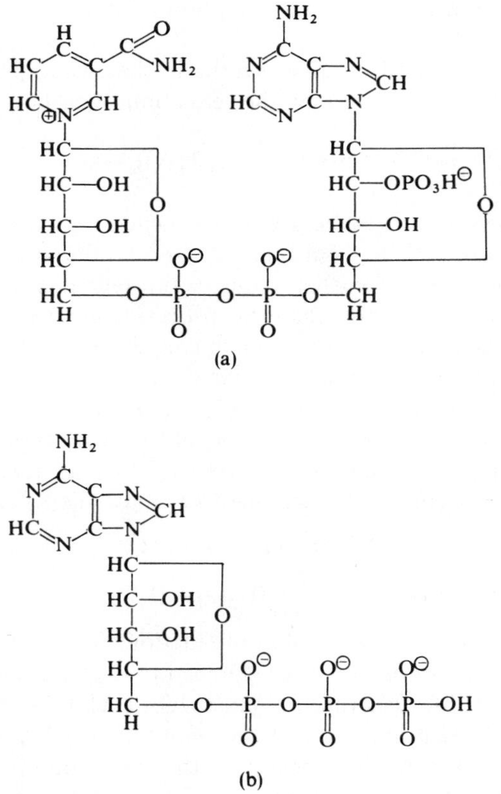

(a)

(b)

Fig. 8.6   Formulae of (a) nicotinamide adenine dinucleotide phosphate (NADP) and (b) adenosine triphosphate (ATP).

processes by the nonstoicheiometric equations

$$\text{ADP} + \text{inorganic phosphate} \xrightarrow{h\nu} \text{ATP} \qquad (8.46)$$

$$\text{NADP} + \text{H}_2\text{O} \xrightarrow{h\nu} \text{NADPH} + \text{O}_2 \qquad (8.47)$$

It is well known that ATP is an 'energy-rich' phosphate used in many biochemical processes; the storage of chemical energy arises from the energy available in hydrolysis of ATP to ADP and H_3PO_4 (about $25\,\text{kJ mol}^{-1}$). Since reaction (8.46) can occur independently of CO_2 reduction, and in an anaerobic environment, it seems possible that the original development of the use of light by organisms was primarily for the storage of energy rather than for the synthesis of new organic matter. The development of photosynthesis proper would then have been a later evolutionary step.

The trapping and use of solar radiation depends on the presence of chlorophyll in the plant. Figure 8.7 shows the structure of the most ubiquitous chlorophyll, chlorophyll-a. The resonance of the conjugated system brings the optical absorption into the visible region of the spectrum, at wavelengths where the solar intensity is highest at ground level. At the same time, the stability conferred by the porphyrin structure ensures that absorption of radiation is followed by energy transfer or radiative processes rather than by dissociation of the chlorophyll; chlorophyll is an exceptionally efficient photosensitizer because of its ability to trap the energy of radiation and pass it on from one molecule to another until conditions are favourable for the sensitized reaction. In organic solutions the fluorescence yield is about 0.3 (although in the natural state it is much less), which is further evidence for the stability of the molecule.

The absorption spectrum of chlorophyll-a in organic solvents shows two major and two minor absorption peaks. One of the major peaks lies in the blue-violet and the other in the red region of the spectrum. In photosynthetic organisms, the chlorophyll-a is usually accompanied by one or more auxiliary pigments whose function may be to absorb radiation at wavelengths between the chlorophyll-a peaks; chlorophyll-b (chlorophyll-a with the 3-methyl group replaced by—CHO) and carotenes are probably the most important absorbing auxiliary pigments in the higher plants. The auxiliary pigments appear always to transfer their excitation energy to chlorophyll-a. Red fluorescence ($\lambda_{\text{max}} \sim 680\,\text{nm}$) of chlorophyll-a alone is seen in mixtures of it with auxiliary pigments, even though the latter substances absorb the incident radiation; this same red fluorescence is also observed on irradiating chlorophyll-a in its blue-violet absorption region. Thus, the maximum amount of energy available in excitation of a single chlorophyll molecule is not more than $180\,\text{kJ mol}^{-1}$. Photophosphorylation (reaction 8.46) is therefore

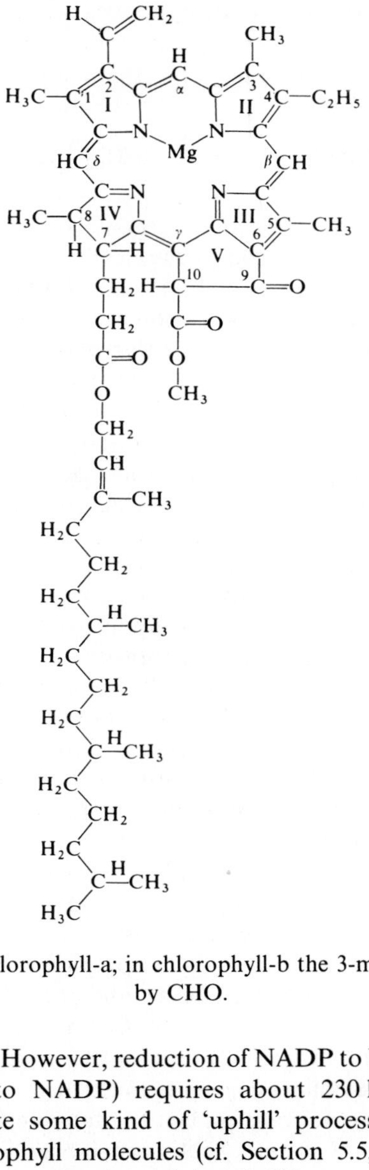

Fig. 8.7 Formula of chlorophyll-a; in chlorophyll-b the 3-methyl group is replaced by CHO.

energetically possible. However, reduction of NADP to NADPH (i.e. electron transfer from H_2O to NADP) requires about 230 kJ mol^{-1}, and it is necessary to postulate some kind of 'uphill' process, possibly involving several excited chlorophyll molecules (cf. Section 5.5, p. 117). The actual energy requirement for reduction of 1 mol CO_2 to carbohydrate is near 470 kJ; with a quantum demand for the overall process of 8, the efficiency of the multiple-quantum processes would be as high as 33%. Some rather special and efficient 'uphill' mechanism must therefore be operating in the photosynthetic plant.

Chlorophyll in the higher living plants is always to be found associated with lipoprotein membranes, which are organized into a highly ordered structure known as the chloroplast. The molecular arrangement in the chloroplast appears to be an essential component of the natural photosynthetic process. The chloroplast functions as a device both for the capture of light, and for the conversion of light energy to chemical energy. Several compounds other than chlorophyll are found in the chloroplast. These include the auxiliary pigments, and carotenoid compounds which can both act as auxiliary pigments and protect chlorophyll against oxidative degradation (cf. p. 151). Also present are quinones (e.g. plastoquinone, α-tocopherol quinone, vitamin K) and proteins known as cytochromes. We shall see later that the quinones and cytochromes play roles in photosynthesis as important as those of the auxiliary pigments and carotenoids.

The membrane systems within the chloroplast seem to consist of a number of flattened sacs, which periodically approach each other closely to form the so-called 'grana' (see Fig. 8.8). Electrons may be transferred from one side of the membrane to the other, in such a direction that oxygen is released on the inside and reduction occurs on the outside. The number of chlorophyll molecules present in each chloroplast is directly related to the number of membrane surfaces, with roughly 10^9 chlorophyll molecules in a typical chloroplast. It seems that the pigment (mainly chlorophyll) molecules might be spread as monolayers on the membrane surfaces, maximizing the surface area of the pigment for light absorption and for energy transfer at specific sites on the membrane. Experiments using flashes of light show that the rate of oxygen production in plants increases with intensity up to a limit corresponding to the excitation of one in every 300 pigment molecules. That result does not mean, however, that the other pigment molecules are always inactive, because the quantum yields measured at low intensities demand that virtually every absorption event contribute to oxygen production. Rather, an

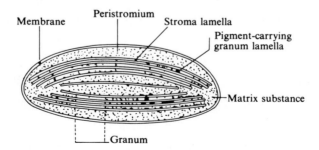

Fig. 8.8 A grana-carrying chloroplast. Other photosynthetic structures are known, but this form is the most important in higher plants. (From Thomas, J. B. (1965). *Primary photoprocesses in biology*, p. 128, Fig. 40. North-Holland, Amsterdam.)

array of pigment molecules can transfer energy to a single reaction centre, only one molecule needing to absorb radiation. At low light intensities, the rate of reaction is determined by the number of arrays excited, whereas at some higher intensity all arrays are active, with one molecule absorbing, and further increases in intensity cannot increase the rate. Energy transfer, via the coulombic long-range mechanism (see Section 5.3), redistributes excitation from the primary absorber to other pigment molecules and ultimately to the reaction centre. In this elegant way, the available light is collected efficiently without molecules of all components of the subsequent chemical system having to be present for each individual light absorber. The complex array of pigment molecules thus functions as a *light-gathering antenna*.

Efficient photosynthesis appears to require the simultaneous excitation of more than one photosynthetically active pigment, a result that suggests the possibility of two major processes in the energy-conversion reaction of photosynthesis. The quantum efficiency of photosynthesis drops at wavelengths longer than the red absorption maximum (the Emerson or 'red-drop' effect), even though absorption in this region (675–720 nm) should still populate $S_1^{v=0}$ of the chlorophyll-a. However, if supplementary light of shorter wavelength ($\lambda < 670$ nm) is added to the irradiating beam, the quantum yield for photosynthesis is considerably increased. The low quantum yields obtained with long-wave radiation can, indeed, be restored to

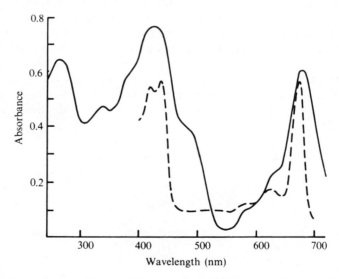

Fig. 8.9 Absorption spectrum of *Euglena gracilis* chloroplast *in situ* (obtained by microspectrophotometry) (———); absorption spectrum of purified chlorophyll-a in acetone (– – –). (From Wolken, J. J. (1986). *Light and life processes*, Chapter 8. Van Nostrand Reinhold, New York.)

the 'normal' values by simultaneous illumination with short-wavelength radiation.

Figure 8.9 shows that the long-wavelength chlorophyll absorption in a chloroplast is shifted to the red, probably partly as a result of complex formation with proteins. More detailed examination of absorption spectra reveals at least two forms of chlorophyll in the chloroplast, which may possibly be chlorophyll-a complexed to different proteins, or present as monomers and dimers. In the absence of more specific information about the two forms of chlorophyll, they are known as 'pigment systems 1 and 2' (PS1 and PS2), or P_{700} and P_{680}, the subscripts indicating the wavelength of absorption. PS2 absorbs at shorter wavelengths than PS1, probably because absorption occurs in an auxiliary pigment (e.g. chlorophyll-b in green plants): fluorescence studies indicate, however, that the excitation always resides on the chlorophyll-a and not on the auxiliary pigment.

A scheme suggested by Hill and Bendall for the formation of ATP and NADPH is illustrated in Fig. 8.10. This figure does not describe the mechanism of the phosphorylation and reduction, but rather shows the energetics of the processes in redox potential form. The scale is arranged with negative

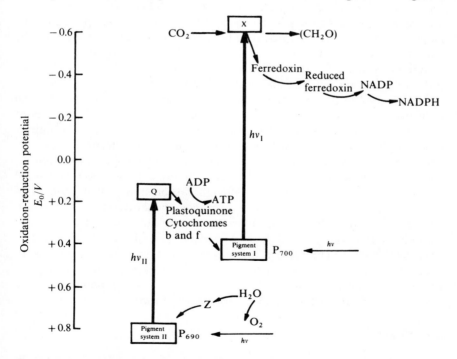

Fig. 8.10 The 'Z' scheme of photosynthesis: the two pigment systems in photosynthesis, pigment system I and pigment system II, and the bridge connecting them.

potentials (strongly reducing couples) at the top and positive potentials (strongly oxidizing couples) at the bottom. Electron transfers from the reducing species of one couple to the oxidizing species of a couple below it are, therefore, spontaneous. In outline, the photosynthetic scheme consists of two partial oxidation and reduction reactions, electrons being transferred against the potential gradient by the absorption of light. Absorption of light by PS2 leads to the formation of a strong oxidant, which eventually yields molecular oxygen and a weak reductant. The photoprocess driven by PS1 generates a strong reductant and, concomitantly, a weak oxidant. The weak reductant of the PS2 reactions and the weak oxidant of the PS1 reactions are linked by a chemical 'bridge' of compounds involving plastoquinone and cytochromes. NADPH is derived from the strong reductant finally produced, and ATP is generated in the bridge segment.

Known redox potentials indicate possible identities for several of the oxidants and reductants; some reasonable compounds are indicated in Fig. 8.10. The exact nature of the species X, Q, and Z is not known, although there is strong evidence that Z involves a manganese ion in some form. The initial process effects the generation of O_2 from water by the oxidation of hydroxyl ions, and the first electron acceptor is plastoquinone, which acts via the 'bridge' as a donor to the PS2 redox system; the starting compound of this latter system is probably cytochrome-f. Recent evidence suggests that the strong reductant produced may be the reduced form of ferredoxin, an iron-containing protein found in the cytoplasm which has the relatively high negative reduction potential of -0.43 V in this system.

The function of the chlorophyll in photosynthesis is now clear: under the influence of light it can cause electron transfer and bring about oxidation–reduction changes. Two kinetic processes are of particular importance in photosynthesis—electronic energy transfer and electron transfer—and each has associated with it a pigment–protein complex. True photochemical processes that involve electronically excited states are completed within about the first nanosecond after the absorption of light, and time-resolved experiments are now able to investigate the shortest times of interest. The spectroscopic information obtained is complex, and the next stage in exploring photosynthesis will depend on separation of the pigment–protein complexes from the chloroplast. A few of the components have been crystallized, and it seems likely that a detailed description of the primary step in the complex molecular aggregate will soon further our understanding of what is one of the most important of all photochemical processes.

8.4 Vision

Three phyla (arthropods, molluscs, and vertebrates) have developed well-formed eyes, although the anatomy and evolutionary development of vision in the three phyla are entirely different. It is therefore remarkable that the photochemistry of the visual process is nearly identical for the three types of eye. Vision is stimulated in each case by the photochemical transformation of a pigment containing a moiety related to vitamin A (retinol)[†]. It is this photochemistry with which we are mainly concerned, although the photo-receptive structures of the eye must be discussed in so far as they affect the photochemistry.

The gross anatomy of the vertebrate eye—in particular, the system of lens and retina—is too well known to need description here. The receptors of the retina consist of 'rods' and 'cones': the former possess high sensitivity and are used at low light intensities, while the latter are less sensitive but may carry colour-selectivity. Electron microscopy has revealed the structure of the rods and cones for some species, and Fig. 8.12 is a diagram of the rod and cone outer segments of the *Necturus* eye. A number of lamellae are formed by the

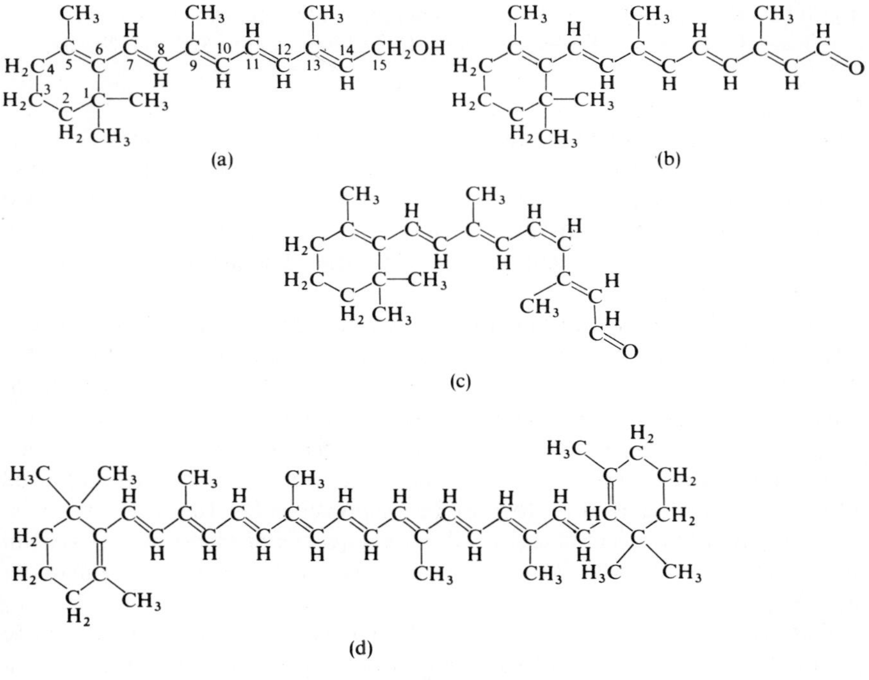

(a)

(b)

(c)

(d)

Fig. 8.11 Formulae of (a) retinol, (b) all-*trans*-retinal, (c) 11-*cis*-retinal, (d) β-carotene.

[†] There are two vitamin A compounds, A_1 and A_2; A_2 contains a second double bond at the 3:4 position in the ring (see Fig. 8.11). In this section, 'vitamin A' refers to A_1; A_2 may be important in the visual processes of some fish.

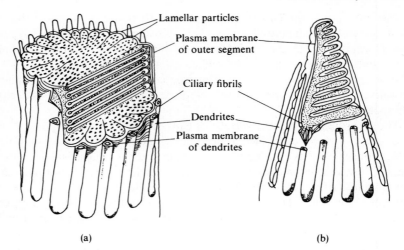

Lamellar particles

Plasma membrane of outer segment

Ciliary fibrils

Dendrites

Plasma membrane of dendrites

(a)   (b)

Fig. 8.12   Sketch of (a) rod and (b) cone outer segments of the eye of *Necturus*. (From Wald, C., Brown, P. K., and Gibbons, I. R. (1963). *J. Opt. Soc. Am.* **53**, 20.)

infolding of plasma membranes, and these lamellae are the carriers of the visual pigments. Observations of the rod outer segments of the frog have shown that illumination transforms the straight cylinders into crumpled structures with many transverse 'breaks', as if the lamellar discs had fallen apart. The effect is consistent with a light-induced structural change in the visual pigment.

In 1876 Böll discovered that the rose colour of the frog's retina faded in bright light. This bleaching of the so-called 'visual purple' demonstrated explicitly the occurrence of a photochemical reaction in vision. Subsequent studies showed that the bleaching is reversible if the retina is kept *in situ*: the reversibility was lost in solutions of visual purple extracted from the retina, even though the initial photobleaching still occurred. It is now recognized that the bleaching is too slow to be responsible for the sensory visual response, and that it is the end result of a sequence of reactions involved in nerve excitation. We now turn to an examination of the nature of the pigment and its photochemistry.

Chemical studies of extracts from the retina show the visual pigments to be compounds of a carotenoid substance with a protein. Rhodopsin (visual purple), which is typical of the pigments, contains 11-*cis*-retinal as the carotenoid chromophore, and the protein scotopsin. Figure 8.11 shows the relationship of retinal to retinol (vitamin A) and to β-carotene. Animals derive their retinol from carotenoids of plant origin, and retinal is produced in the retina by enzymic oxidation of retinol. Scotopsin is the class of pigment protein found exclusively in rods (*photopsin* is found in cones, and when bonded to retinal forms the pigment *iodopsin*). The opsins are proteins with relative molecular masses of about 40 000. Rhodopsin (bovine or ovine) has 348 amino acid residues grouped into seven mainly hydrophobic segments that pass between the two sides of the photoreceptor membrane. The bond between the protein backbone and retinal results from condensation of the aldehyde with the ε-amino group in a lysine residue towards the end of the chain (position 296 or its equivalent) to form a Schiff base,

$$C_{19}H_{27}HC=O + RNH_2 \rightarrow C_{19}H_{27}HC=NR + H_2O \qquad (8.48)$$

which is probably protonated. The photophysiological properties of rhodopsin depend on the complexing between chromophore and protein. First, the optical absorption in the red ($\lambda_{max} \simeq 500$ nm for rhodopsin and $\simeq 562$ nm for iodopsin) derives from a red-shifted transition in the 11-*cis*-retinal (λ_{max} in ethanol $\simeq 378$ nm). Secondly, the colour changes (these include bleaching; see below) observed on illumination of rhodopsin do not occur with retinal alone. Thirdly, carotenoids are unlikely to initiate nervous stimulatory responses.

Irradiation of rhodopsin leads to a series of conformational changes that are reflected in the appearance and disappearance of various intermediates of differing colour. The processes may be described by the sequence

$$
\begin{array}{lll}
\text{rhodopsin} & \xrightarrow{h\nu} \text{bathorhodopsin} & \xrightarrow{\text{thermal}} \text{lumirhodopsin} \\
\text{(red)} & \text{(red)} & \text{(orange-red)} \\
& & \downarrow \text{thermal} \\
\text{retinal + opsin} & \xleftarrow{\text{thermal}} \text{metarhodopsin II} & \xleftarrow{\text{thermal}} \text{metarhodopsin I} \\
\text{(colourless)} & \text{(yellow)} & \text{(orange)}
\end{array}
$$

$$(8.49)$$

The first step to form bathorhodopsin occurs on a time-scale of tens of picoseconds, and each subsequent step is 10^2-10^3 times slower than its predecessor. Current opinion ascribes the changes to the inability of the straight all-*trans*-retinal to be sterically accommodated on the surface of the opsin, only the bent 11-*cis*-retinal 'fitting' into the protein. Absorption of a photon leads to photoisomerization, and thus to strained structures and ultimately to cleavage of the protein–chromophore bond. The transition to bathorhodopsin involves isomerization to a nearly all-*trans*- form, but one that has not yet adopted its lowest energy geometry. The more relaxed all-*trans*-isomer appears at the lumirhodopsin stage. At each step the protein backbone rearranges, with prominent changes in one or more buried carboxyl groups becoming apparent in metarhodopsin I. Formation of metarhodopsin II involves deprotonation of the Schiff base, as well as significant changes in the lipid structure. It is the metarhodopsin II that triggers the next set of biochemical steps, to which we shall return shortly. The changes in optical absorption seem consistent with the picture presented. Lowering of the energy of the excited state due to interaction of retinal with opsin leads to a red shift, and the stronger the interaction, the larger the shift. As the progressively more strained structures of lumirhodopsin and the metarhodopsins are formed, the shift becomes smaller and the absorption maximum moves towards the blue. In the case of bathorhodopsin itself, which absorbs at slightly longer wavelengths than rhodopsin, it may be that the ground state lies at higher energy than the starting rhodopsin as a result of the strained geometry. The cycle is completed by the slow thermal isomerization of all-*trans*-retinal to the 11-*cis*-isomer that combines spontaneously with opsin, and additional retinal can be supplied from vitamin A as needed.

The electronic states involved in the *cis–trans* isomerization have not been established unequivocally. Although a transient spectrum consistent with a triplet species is observed on the flash photolysis of all-*trans*-retinal, there is no evidence for triplet states in the isomerization of rhodopsin. It is possible that the $^3(n, \pi^*)$ levels lie above the excited state (perhaps a π, π^* state) in the protonated Schiff base, but below it in the free aldehyde, so that triplets are only formed in free retinal. It seems likely that, in the photoexcited retinal chromophore of rhodopsin, there is a transfer of positive charge from the Schiff base to the conjugated π-system of the retinal.

Colour vision is associated with the cones rather than with the rods. We have already seen that the pigment iodopsin has its absorption maximum at slightly longer wavelengths than the maximum of the absorption of the rhodopsin belonging to the rods. The cones are less sensitive than the rods, and the spectral response of the eye shifts towards the red in going from dim to bright light, as expected. Vertebrates appear to perceive colour through the operation of a three-colour system. Three different cone pigments seem to be implicated,with absorptions in the blue, the green, and the red wavelength regions. Although microspectroscopy shows the presence of different pigments, their isolation has proved elusive. It seems likely that the pigments are very closely related to rod rhodopsin. One approach to understanding the protein structure has been to study the genetic DNA that codes for the pigments, and hence to derive probable amino acid sequences. Charged amino acids near the retinal π-system alter the energies of ground and excited states, and the deduced structures of the cone pigments are consistent with a model of the retinal absorption spectrum being tuned by interaction with neighbouring charged amino acids. Each cone possesses only one pigment, and so has a specific spectral response. The responses from the cones might be used individually or in a pairwise fashion to allow the brain to interpret the colour image.

We now turn to a brief consideration of how the photochemical changes described so far become converted to an electrical impulse that stimulates the brain. There is evidence that a single quantum of radiation can stimulate a retinal rod. The absorption of one quantum does not, however, result in vision, and several quanta (between two and six is considered a reasonable estimate) must reach the same rod within a relatively short period. Even so, the process is remarkably efficient, and the energy of the ultimate reaction greatly exceeds that absorbed by the visual pigment. The absorption of light appears to initiate a chain reaction that derives its energy from metabolism, and visual excitation is a result of 'amplification' of the light signal received at the retina. The photoreceptor is the biological equivalent of a photo-multiplier tube, which converts photons to an electrical signal with high gain and low noise (see Chapter 7). Both photoreceptor and photomultiplier achieve high gain in a cascade of amplifying stages. Visual pigments are integral membrane proteins that reside in the plasma and disc membranes of the photoreceptor outer segment. Photoisomerization of retinal triggers a series of conformational changes in the attached protein that create or unveil an enzyme site. A cascade of enzymatic reactions follows, which ultimately produces a neural signal. The electrical response starts with a transient hyperpolarization due to closure of several hundred Na^+ channels in the plasma membrane. Thus, a messenger molecule transmits information from the receptor disc to the plasma membrane. A likely candidate for the messenger is the strained, energy-rich, cyclic phosphate cGMP (guanosine-3',5'-cyclomonophosphate), perhaps in combination with Ca^{2+} ions. Cationic conductance of both rod and cone plasma membrane has been shown to be controlled directly by cGMP. The light-induced structural changes in the disc thus activate the transduction mechanism that itself generates a potential spreading over the plasma membrane. Details of the transduction and amplification mechanisms are still being established. One suggested scheme concentrates on the pivotal importance of phosphodiesterase (PDE) in controlling the concentration of cGMP. In this scheme, metarhodopsin II is able to interact with a protein, transducin, and induce a rapid exchange of GTP for GDP bound on the transducin (GDP and GTP are the guanosine analogues of ADP and ATP discussed in the last section). Several hundred transducin molecules can be activated by each photoactivated rhodopsin molecule, so that the first stage of amplification may arise here. The GTP–transducin complex is the activator of PDE, which controls cGMP concentrations and hence electrical activity. One activated PDE molecule can hydrolyse approximately 1000 cGMP molecules per second to introduce a second stage of signal amplification.

Visual images can be recorded by photochemical methods, and our examination of how man has harnessed photochemistry in his service begins in the next section with a discussion of photoimaging.

8.5 Photoimaging

The making of a more or less permanent record of light and shade by photography represents the best-known of all applied photochemical processes. Photography is one of a series of *photoimaging* techniques in which photons are used for the capture and replication of image information. Besides photography, other obvious large-scale applications of imaging include office copying and the preparation of various kinds of printing plates. If the image modifies the properties (e.g. the solubility) of a material used to protect some underlying medium, then subsequent treatment may allow the image to be transferred to the formerly protected surface. Such materials are known as *photoresists*, and are of enormous importance in the production of printing plates, integrated circuits and printed circuit boards for the electronics industries, the manufacture of small components such as electric razor foils and camera shutter blades, and many other applications besides. Considerable interest now attaches to photoimaging as a route to all-optical, as distinct from magnetic, memory storage devices in which writing and reading of the stored elements of information is achieved by electromagnetic radiation in the photochemically active infrared to ultraviolet part of the spectrum. Applications of optical reading to video and audio ('compact-disc') technologies are now well established, and there are enormous advantages to be reaped in optical read–write memories for computers.

Photoimaging can be divided into the three stages of capture, rendition, and readout. These stages are illustrated in Fig. 8.13 for a typical imaging system, and several end-products are indicated. The capture process in the example is photopolymerization; description of this important technique is deferred until Section 8.8.1. In general, image capture is the photochemical step, with image rendition consisting of the subsequent thermal reactions. Image readout then requires the development of the rendered image into a form that differentiates it physically from the unexposed background. This step can involve changes in optical properties such as opacity, scattering power, or refractive index; increases or decreases in solubility; or alterations in wettability, tackiness, adhesion, etc. A few examples of the creation of a desired product must suffice to illustrate the myriad applications that exist. Optical changes form the basis of ordinary photography; solubility changes can be used to generate three-dimensional relief patterns for printed circuits, printing plates, or three-dimensional topographic maps; wettability changes are used to produce lithographic printing plates, and enhanced tackiness can be used to pick up pigments to yield pigment-toned images in printing.

By far the most common form of the photographic process (both for monochrome and for colour photography) is based on silver halides as the photosensitive materials, and the principles will be explained in the next section. However, as an introduction to the next few sections, we consider

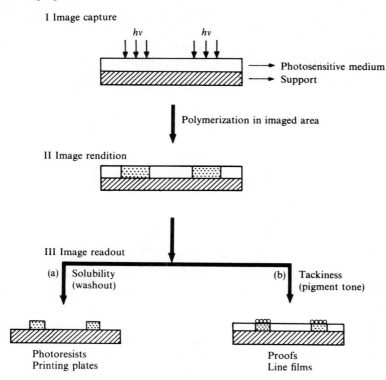

Fig. 8.13 Typical elements in photoimaging. (From Eaton, D. C. (1986). Dye sensitized polymerization, *Adv. Photochem.* **13**, 427)

some 'unconventional' systems that illustrate some of the general features set out in the last paragraph. Motivations for the use of non-silver-halide systems include the shortage and cost of silver compounds, the ever-increasing need for dry processing, and the desirability of direct and immediate access to the recorded information. In addition, silver halide photography depends on the formation of silver particles, and the ultimate resolution is limited by the fineness of the grain size. Some of the unconventional systems produce images by altering individual molecules, so that the potential resolution is far higher, although it is achieved usually at the expense of greatly reduced sensitivity to light.

The examples that follow are all based on the photodecomposition of aryl diazonium salts, $ArN_2^+ X^-$. Ultraviolet radiation photolyses these salts via their excited singlet states

$$ArN_2^+ X^- + h\nu \rightarrow Ar^+ + N_2 + X^- \tag{8.50}$$

although spectral sensitization by dyes (cf. Section 8.6) can also allow visible light to initiate the decomposition. In the *diazotype* process, the diazonium salt is present in a coating on some support (often paper). Those parts of the material that are exposed to light suffer decomposition of the salt. If subsequently the sensitive material is 'developed' with a phenolic or amino coupler, then an azo dye can be formed by the reaction

$$
\text{ArN}_2^+\text{X}^- \quad + \quad \text{(naphthol-OH)} \quad \xrightarrow{\text{Base}} \quad \text{(naphthol with N}_2\text{Ar and OH)} \quad + \quad \text{HX} \qquad (8.51)
$$

only in those areas that have not been exposed to the light. A positive image is thus produced, whose colour depends on the choice of coupler.

One physically-developed image system relies on the production of N_2 in reaction (8.50). The diazonium salt is dissolved in a thermoplastic layer that is heat-treated after exposure to light. Trapped nitrogen is expanded by the heat and forms small bubbles, about 1 μm in size, in the softened plastic. These 'vesicles' scatter light efficiently, so that exposed areas appear opaque in transmitted light. The *vesicular* process therefore yields a negative image.

Some *relief* processes are also based on diazonium salts. For example, photodecomposition of the salt contained in a solvent-soluble polymer can be made to reduce the solubility of exposed areas. Subsequent solvent treatment thus dissolves unexposed regions preferentially, and leaves a negative relief image that can be used typically in printing operations.

Having established the principles of image formation, we turn now to some further important practical implementations, starting with silver halide photography.

8.6 Photography

In silver halide photography, microcrystalline grains of silver halide suspended in gelatin are coated onto a suitable support (film, glass plate, paper, etc.) to form the light-sensitive 'emulsion'. Prolonged exposure to light causes darkening of the emulsion—the *print-out effect*—which X-ray powder patterns show clearly to be a result of metallic silver formation. Much shorter exposures produce a so-called *latent image* in the silver halide grains. This latent image may be turned into a visible silver deposit by a 'developer', which is a suitable reducing agent. All developers are, in fact, thermodynamically reducing towards silver halides, and the presence of the latent image seems to lead to an increased rate of reduction to metallic silver rather than to a change in the ultimate reducibility of the emulsion. Extended development of unexposed emulsion leads finally to darkening ('fog'), so that the discrimination by development between exposed and unexposed areas depends on the difference in reduction rate for the two areas.[†]

Several experiments give direct evidence that the latent image consists of metallic silver present in the halide grains, but at much lower concentrations than in the print-out image. Using techniques capable of detecting optical density changes around 10^{-6}, it is possible to find measurable silver densities in latent images even at the threshold of developable exposures. There is also a marked similarity in the influence of environmental factors (e.g. the presence of electric fields or of crystal imperfections—see below) on the location of print-out silver particles and of development centres. Our discussion of the primary photochemical processes will therefore deal mainly with the production of print-out silver, and it will be assumed that latent image formation is photochemically identical but that it involves much lower conversions. However, one important feature of latent image formation is the decrease in emulsion sensitivity at very low light intensities ('Reciprocity-law failure'), which indicates the presence of a multi-quantum process. There is evidence that a single silver atom is generally unstable in the halide lattice, having a lifetime of only a few seconds, and that at least two atoms are needed unless there is some pre-existing stabilizing centre.

The presence of metallic silver atoms in the latent image appears to lower the activation energy for the reduction reaction in development, and thus enhance the rate. Development, once it has been initiated at one site on a grain, proceeds with an increasing rate as more and more silver is formed, until the entire grain is developed. This autocatalytic activity of silver has been clearly demonstrated: a low concentration (about 10^{15} atoms cm^{-2}) of metallic silver evaporated onto a silver halide surface renders the surface 'developable'.

Silver halides show the phenomenon of photoconductivity, and it is believed that irradiation of silver halide raises photoelectrons from the valence band to the conduction band (see p. 236) of the halide. The mechanism for the production of free silver then involves the migration of the photoelectrons and of interstitial silver ions to preferential sites in the halide; free silver atoms are formed by the combination of silver ions and electrons. The free silver so formed acts as an efficient trap for photoelectrons produced subsequently, so that further silver ions are discharged *near the same place* as the initial atom. Specks of silver grow, therefore, at the original preferential site. The positive 'holes' left behind by the electrons may have some mobility, and they can diffuse towards the surface of the halide grain to liberate free halogen. Figure 8.14 represents the mechanism, and is based on a form first described by Gurney and Mott. An alternative scheme, proposed by Mitchell, involves

[†]In normal practice the developed image corresponding to exposed and unexposed areas is rendered permanent by 'fixing': the unexposed (and, hence, unaffected) grains of silver halide are dissolved in sodium thiosulphate solution immediately after development.

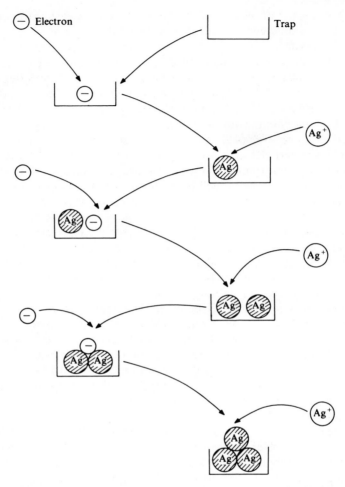

Fig. 8.14 Representation of silver image formation based on the Gurney–Mott concentration principle. In Mitchell's alternative scheme, the first act is trapping of an electron by Ag^+; the Ag formed adsorbs a second Ag^+ ion, and so on.

initial trapping of an electron by an Ag^+ ion, and subsequent adsorption of Ag^+ on the growing silver speck to trap further electrons. In either case the basic processes are similar. The steps up to the formation of the two-atom speck are reversible, which is consistent with the experimental evidence for latent image stability only with aggregates of more than two atoms (see above).

The hypothetical Gurney–Mott (or Mitchell) mechanism has been well-substantiated by experiment. The photoconductivity of silver halides *previously darkened by irradiation* is less than that of undarkened halides, which indicates that the colloidal silver particles (or the physical imperfections introduced into the lattice by their formation) are effective in trapping electrons. The participation of charged species in image formation has been shown in an experiment in which a crystal of silver chloride is held between two electrodes, and exposed to radiation through a semi-transparent conducting aperture in one electrode. Strongly absorbed light is used, and in the absence of an electric field, silver image formation is restricted to a region near the crystal surface. If, however, a strong electric field is applied,[†] with the illuminated electrode made negative, then photoelectrons are displaced towards the interior of the crystal and the region of darkening is similarly displaced, which indicates that photolytic silver separates at sites where the photoelectrons are trapped. The same effects have subsequently been shown in the microscopic grains of a silver bromide emulsion using *weakly* absorbed light. Electron microscopic examination shows that, in the absence of an applied field, silver grains are liberated uniformly throughout the grain. In the presence of the field, silver particles concentrate almost entirely near the positive side of the grain. Furthermore, a high concentration of atomic bromine builds up near the negative side, thus demonstrating the drift of 'holes' in the grain and confirming the photoliberation of halogen.

The escape of halogen is necessary for photochemical change; unless holes diffuse to a grain surface and liberate halogen, they recombine with electrons and no free silver is formed. Investigation of the silver distribution in irradiated crystals of silver halide has shown that the silver is concentrated in a subsurface layer which is at most a few microns thick. Further, the quantum yield for photodecomposition in large crystals of silver bromide is small when the exposing light is weakly absorbed (e.g. $\phi \sim 0.02$ at $\lambda = 436$ nm), but increases to near unity at very short wavelengths at which all the light is absorbed near the surface. It has been shown that with weakly absorbed light all liberated bromine comes from a surface layer about 0.3 μm thick, and that the quantum yield is high in that surface region.

A real crystal or grain of silver halide may possess chemical impurities and physical imperfections; in the case of emulsions, the gelatin may also enter into the photochemical processes.

The role of chemical impurities is shown clearly by experiments on silver halides with substitutional cuprous ions. The quantum yield for silver production is near unity even in the centre of a crystal, and the number of silver atoms formed at saturation level is equal to the number of cuprous ions initially present. EPR spectra show the formation of cupric ions concomitant with silver atom production, and the impurity cuprous ions appear to act,

---

[†] The actual experiment involved flashes of light and a pulsed electric field to prevent significant electrolysis of silver chloride.

therefore, as traps for the holes in the bulk of the crystal. Flash photolysis studies confirm this interpretation. A photo-darkening is observed in both pure and cuprous-doped silver halide crystals, and the rate of build-up is similar. However, the darkening is transient in the pure crystals, and fades in a few milliseconds, whereas it is stable in the doped silver halide.

Extended crystal imperfections serve two functions in photographic image formation. First, they provide charged centres which act as traps for electrons and holes, and, secondly, they provide sites from which interstitial silver ions are readily generated. It is probable that in the volume of a halide grain the principal sites for silver separation are on internal imperfections such as jogs on edge dislocation lines, grain boundaries, and twin planes (although at the surface—which can be considered as an imperfection itself—there is no shortage of electron traps, and additional imperfections are probably unnecessary). The lifetime of photoelectrons is increased from about 1 to 10 μs by annealing out physical imperfections. Again, introduction of slip planes by mechanical deformation leads to preferential darkening of the deformed regions. Microscopic examination of print-out silver shows that dislocations and mosaic boundaries are effectively 'decorated' by the silver; essentially similar decoration of imperfections is seen after development of a latent image.

Gelatins containing labile sulphur or reducing groups have long been known to increase the sensitivity of a photographic emulsion, and in present-day manufacturing techniques controlled additions of sensitizers are made to inert gelatins. The exact mode of action of chemical sensitizers is not yet established, although it seems certain that silver sulphide is formed in sulphur-containing emulsions. The sulphide can then act at the sites of image centres, either to provide deeper electron traps or to increase stability during the earliest stages of image formation. Silver sulphide can also reduce recombination of electrons and holes and remove bromine, since it can capture holes or bromine.

Spectral sensitization of silver halide emulsions can be achieved by adsorption of suitable dyes onto the halide grains. Such sensitization is important, since it permits image formation by radiation of wavelength longer than that effective with unsensitized emulsions (say 490 nm—blue-green—for silver iodide emulsion), and it offers an excellent example of a reaction photosensitized by energy or electron transfer. Indeed, spectral sensitization of photographic emulsions seems to have been the first recognized case of photosensitization (1873).

Perhaps the most important class of sensitizing dye is that of the cyanines: these dyes contain heterocyclic or benzenoid nuclei joined by a $=CH(-CH=CH)_n$ bridge, the π electrons of which take part in the spectral transitions leading to sensitization. It is a characteristic of these dyes to be strongly adsorbed on to the silver halide grains. The fluorescence yield of the adsorbed dyes is much lower than that of ordinary solutions. However, the phosphorescence yield is also small, and the decrease in fluorescence does not appear to result from a rate of $S_1 \rightsquigarrow T_1$ ISC enhanced by the heavy-atom effect (cf. p. 89). Rather, the results suggest that the fluorescence is quenched by transfer to excitation from the S_1 level of the dye to the silver halide. The close correspondence between dye absorption spectra and the spectral sensitivities of sensitized emulsions indicates clearly that the excitation transfer is also responsible for sensitization of the photographic process.

Conduction electrons appear in the silver halide crystal when the dye absorbs radiation, just as when the silver halide itself is the absorber. Photoconductivity of thin dyed crystals of silver halide is observed at wavelengths absorbed by the dye, and longer than those to which the undyed crystal responds. Sensitization could be a result of transfer from the dye to the halide either of an electron or of excitation energy. The high efficiency of sensitization by many dyes points to the electron transfer process as the

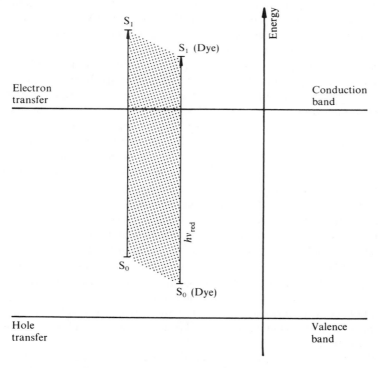

Fig. 8.15 Schematic representation of the ground and first excited singlet levels of one class of dye in spectral sensitization. The energy levels of the dye are shown in relation to the top of the valence band and the bottom of the conduction band of silver halide.

dominant mechanism, at least with the efficient dyes. Figure 8.15 shows the process schematically. The long-wavelength limit of the unsensitized process (ca. 490 nm, equivalent to 247 kJ mol^{-1}) probably corresponds to the minimum energy gap between valence and conduction bands. However, dye-sensitized image formation can be achieved at wavelengths as great as 1300 nm (101 kJ mol^{-1}), at which the energy is insufficient to excite an electron in the halide. Energy-rich surface sites and perturbations by the neighbouring silver halide ions play an important part, because they determine the relative energy by which the S_0 level of the dye lies above the valence band in the silver halide. The range of energies shown in the figure represents the statistical spread. A restriction is placed on the dye energy levels, since S_1 must lie either above the bottom of the conduction band or close enough to the band that thermal excitation can effect the electron transfer.

Although regeneration of the dye molecule is not necessary for the sensitization of latent image formation, it is, in fact, observed for print-out image production, and any mechanism proposed for sensitization must account for dye regeneration. The hole created in the dye molecule must be replaced. Transfer from the valence band of the regular lattice would require an activation energy *at least* equal to the separation between the top of the valence band and S_0 of the dye. Such a process would be improbable at room temperature for dyes providing long-wavelength sensitization, and it seems likely that regeneration involves passage of an electron from an energy-rich surface halide ion.

The temperature dependence of the excitation transfer process can be measured by exposing an emulsion at reduced temperature, and subsequently developing under normal conditions. There is a drop in sensitivity that must be ascribed to a reduction in the efficiency of electron transfer. Marcus–Levich electron-transfer theory predicts satisfactorily the observed variation of sensitivity with temperature in many cases.

Another use of dyes in silver halide photography is in producing images for colour photography. Here, much larger quantities of dye are used than in sensitization, and the dye becomes incorporated in the final image. A three-colour 'subtractive' process is used to give the entire range of visible hues. Three layers of emulsion are used, each being spectrally sensitized to a different spectral region (blue, green, and red), and possessing integral filter dyes if necessary. Exposure to a coloured image thus leads to the formation of different latent (silver) images in each layer. Although development involves reduction of the silver, as with black and white photography, in colour photography it is the consequent oxidation of the developer that is of interest. The oxidized developer reacts with a 'colour coupler', often incorporated in the emulsion layer, to produce an image dye. A typical system involves a derivative of benzene-1,4-diamine as developer, and formation of a magenta image is exemplified by the reaction

$$\text{(8.52)}$$

Developer   Coupler   Magenta dye

We note that since the silver image formation is a negative process (dark areas produced by exposure to light), it is also necessary to arrange for the 'negative' or subtractive colours to appear on development. The magenta image formed in reaction (8.52) is a response to exposure to green light; similarly, a cyan dye is produced by red light, and a yellow one by blue light. A second negative colour process, such as printing on paper, then yields a positive reproduction of the original subject with the correct colours.

8.7 Photochromism

Silver halide photography involves the production of an essentially permanent optical effect by means of an irreversible photochemical process. The production of a *reversible* photoinduced colour change is referred to as *photochromism*. In photochromic systems, irradiation drastically alters the absorption spectrum; but when the irradiation source is removed, the system reverts to its original state. In some cases the reversal can be brought about by light of a different wavelength. The visible effect often involves the appearance of colour in a previously colourless material, although changes in colour—for example from red to green—are also known.

Numerous applications of photochromic substances have been suggested, and some of these have entered commercial practice. Photochromic sunglasses and spectacles are familiar, and plastics incorporating a photochromic dye have also been used for aircraft windows that darken in bright sunlight but become lighter again under less intense illumination. Various kinds of data storage are possible, including image storage for uses like those of photography. Very high resolution is possible, and the immediate appearance of the image on exposure, without further treatment, is a potentially great advantage over other processes. A rather more frivolous application is in the manufacture of toy dolls that can be 'suntanned': a photochromic dye is used that produces a brown coloration on exposure to sunlight.

The major difficulty in the use of materials incorporating photochromic materials is the rapid 'fatigue' exhibited by most known photochromic substances. Many of the photochromic systems reported are really able to undergo reversal only a limited number of times. Photochromism based on isomerization (see below) offers the best prospect of good fatigue characteristics, since, with alternative systems that involve bond cleavage, a very small lack of reversibility soon leads to chemical decomposition in side-reactions.

The main mechanisms responsible for photochromic behaviour are isomerization, dissociation, and charge-transfer or redox reactions. Many hundreds of specific photochromic substances are known, and a few examples must suffice to illustrate how photochromism arises.

Many aromatic nitro-compounds exhibit photochromic isomerization: the process is believed to involve photoisomerization from the colourless nitro-form to the coloured aci-form:

$$\text{(8.53)}$$

nitro-form   aci-form

$R_1 = H, \quad CH_3, \quad C_6H_5$ etc; $R_2 = $ electron-withdrawing group

(The aci-form must also undergo some dissociation, since it is a strong acid.)

Irradiation of chromium hexacarbonyl in a plastic matrix (ca. 0.1% $Cr(CO)_6$) leads to the formation of a deep yellow colour as a result of photodissociation of the hexacarbonyl. In the plastic, CO cannot escape, and recombination occurs in about 4 h at room temperature:

$$Cr(CO)_6 \underset{\text{dark}}{\overset{\text{light}}{\rightleftharpoons}} Cr(CO)_5 + CO \tag{8.54}$$

Both organic and inorganic charge-transfer or redox photochromic systems are known. A typical reversible photochemical redox reaction occurs in a mixture of mercurous iodide and silver iodide:

$$Hg_2I_2 + 2AgI \underset{\text{dark}}{\overset{\text{light}}{\rightleftharpoons}} 2HgI_2 + 2Ag \tag{8.55}$$

green   yellow   red   black

Heterolytic bond dissociation is responsible for photochromism exhibited by a large number of spiropyran derivatives

$$\text{(8.56)}$$

colourless   blue

Some derivatives of the class of compounds known as fulgides show a particularly interesting type of photochromism. The forward photochromic reaction that is induced by light of wavelength λ_f may be reversed by light of a different wavelength λ_r, as illustrated by reaction (8.57)

$$(8.57)$$

colourless   coloured

I   II

With a suitable choice of substituents, it may be possible to make I insensitive to λ_r and II insensitive to λ_f, and also to make both forms thermally stable. In that case, an optical information storage system can be devised that allows modification of the stored data, since light of one wavelength can be used to write information, and light of another wavelength to erase it. Such a system would be the optical analogue of the magnetic tapes and discs widely used at present, but would have the advantage of enormously increased density of data storage and of speed of access in reading.

8.8 Photochemistry of polymers

8.8.1 *Photopolymerization: imaging*

The many uses of photoresists were described briefly in Section 8.5. One important application is in the manufacture of electronic integrated circuits, where resists are used to define the doped regions on the silicon substrate that will form the resistors, capacitors, diodes, and transistors of the finished circuit, together with the metal conductors joining the components and the insulating and passivating layers. In producing a complex circuit, there may be several tens of successive stages of imaging, followed by etching, doping, or other processing. Each stage must be lined up with an accuracy as good as a few hundred nanometres. Photographic methods are used to provide the required accuracy, although the use of ultraviolet radiation is being supplemented by shorter-wavelength X-rays or electron or ion beams, as even more components are packed into a small space (Very Large Scale Integration, VLSI). Photoresists in use today are generally based on photopolymer systems; those employed in the semiconductor industry are refined versions of photoresists used for making printing plates. Three typical systems will be described in this section.

   Resists in which the areas exposed to light are removed on development
are referred to as *positive working*, since they leave behind the protective
coating where the light was obscured by the transferred pattern. A positive
photoresist might consist of a film-forming phenol–formaldehyde resin,
mixed with a naphthoquinone–diazide compound that is the photosensitive
component. Nitrogen elimination, followed by ring rearrangement and reaction with traces of water present in the atmosphere or the film itself leads to
the formation of an indenecarboxylic acid

$$(8.58)$$

The resin itself is slightly acidic, and is soluble in alkaline solutions, but the
sensitizer is insoluble in water initially, and protects the resin. However, the
indenecarboxylic acid formed on exposure will dissolve in an alkaline developer, so that all the exposed film will dissolve to leave unsolubilized resin in
the unexposed areas.
   Two important *negative working* photoresists depend on *photopolymerization* for their operation, rather than the alteration of polymer solubility
described in the last paragraph. Photopolymerization can be classified
according to whether each increase in relative molecular mass requires its own
photochemical activation step, or whether many (thermal) polymerization
steps follow absorption of a photon.
   Photoinitiated polymerization falls into the second category, and will be
discussed further in Section 8.8.2. *Photocrosslinking* falls into the first category, and involves formation of crosslinks between pre-existing polymer
chains. Since crosslinked polymers are generally insoluble in any solvent,
photocrosslinking might obviously form the basis of a photoresist system,
and since the material that will be left behind after development by a solvent
will now be that in the exposed regions, the resist will be negative working.
One common negative photoresist uses polyisoprene, which may be natural
or synthetic rubber. Preliminary controlled cyclization of the polymer
improves its film-forming properties, but it still contains methylene groups and
some unsaturation. An aromatic diazide, e.g. 2,6-di-(4′-azidobenzal)-4-
methylcyclohexanone, is mixed with the polymer as a photosensitizer. On
irradiation, the sensitizer yields a nitrene, which can react with the polymer in

the two ways indicated in reaction (8.59)

$$
\begin{array}{cc}
\text{N}_3 & \ddot{\text{N}} \\
| & | \\
\text{Ar} \xrightarrow[-\text{N}_2]{nh\nu} \text{Ar} \\
| & | \\
\text{N}_3 & \ddot{\text{N}}
\end{array}
\qquad (8.59)
$$

and thus form crosslinks between the polymer molecules.

Another important photocrosslinking system involves the photocyclo-addition of pendant cinnamate groups incorporated in polymer chains. For example, on irradiation of poly(vinyl cinnamate), the cinnamoyl groups react to form a crosslinked structure,

$$ (8.60) $$

Poly(vinyl cinnamate)   Crosslinked Poly(vinyl cinnamate)

which is insoluble in solvents. Dimerization can take place from either the singlet or the triplet excited states of the cinnamate chromophore. Only the *trans* isomer of the ester dimerizes, but the *cis* isomer is converted to the *trans* isomer by a photoisomerization reaction. Although the cinnamates them-selves only absorb radiation at wavelengths shorter than about 320 nm, triplet sensitizers such as 4,4′-bis(dimethylamino)benzophenone (Michler's ketone) can increase the sensitivity to near-UV and visible light by a factor of several hundred.

Photoinitiated polymerization is also used in imaging and resist processes that involve radical polymerization of species with vinylic or allylic function-ality, thiol-ene systems, or cationic polymerization of epoxides. Since these polymerizations are similar to those used in the ultraviolet curing of inks and coatings, the chemistry is described in the next section.

8.8.2   *Photopolymerization: curing*

Photoinitiation of polymerization has found little application for the bulk production of thermoplastic linear polymers, because satisfactory low-temperature thermal initiators are available. Rather, the major practical uses of photopolymerization are concerned with *in situ* polymerizations of relatively thin films of materials. Apart from the various uses in imaging, photopolymerization of films is extremely valuable in applications ranging from the drying of decorative and protective coatings on a variety of substrates to the rapid and easily-controllable hardening of resins in dentistry. The drying or hardening processes are generally referred to as *curing*, and the photochemical route offers considerable advantages over other methods. For example, many conventional coating techniques employ large quantities of solvents that play no part in the final cured coating, and whose purpose is to reduce the viscosity of the primary material to facilitate the coating operation. There is thus a waste of raw materials and energy, as well as a pollution and fire hazard. Thermal drying of solvent, or initiation of polymerization cure, requires heating of the entire substrate material. If several layers of material, or colours of inks in decorations, are to be used, then several heat–cool cycles may be needed, and most of the heat goes into warming the high thermal capacity substrate rather than the low capacity coating. Not only is energy lost, but sufficient time must be allowed for each coating to dry or cure properly. In contrast, photoinitiation of the cure works directly on the coating, and the substrate need not be affected. With high intensities, the cure can be completed in a fraction of a second, so that many operations can be carried out successively in a continuously operating production line.

Addition polymerization is used almost universally in photoinitiated curing operations, the monomers containing either multiple bonds (especially olefinic double bonds) or strained rings. Most applications of photoinitiated polymerization involve a free-radical mechanism, with the monomer being based on acrylate esters (CH_2=CHCOOR). Acrylate groups are attached to resin types commonly used in coating technology (epoxides, urethanes, and polyesters). Polyfunctional reactive diluents, resulting from the reaction of polyols with acrylic acid, promote a more rapid cure and a more highly crosslinked coating. Commercially successful photoinitiators have generally been aromatic carbonyl compounds, which have absorption spectra that match available ultraviolet sources well. Substituted acetophenones undergo α-cleavage (Norrish Type I: see Section 3.6, p. 52) to yield initiating radicals. Of these compounds, benzoin ethers have been used extensively. Improved performance is obtained by replacing the alpha-hydrogen by an alkoxy group, as in α,α-dimethoxyphenylacetophenone, which is photolysed according to the scheme

$$\text{(8.61)}$$

Initiators notionally derived from benzophenone work by a bimolecular reaction scheme in which radicals are produced by hydrogen abstraction from a suitable donor. A triplet ketone is formed with high efficiency from the excited singlet first populated on absorption of radiation. Hydrogen abstraction then occurs from the triplet state, the efficiency depending in part on whether the triplet is a π, π^* or n, π^* state. Hydrogen donors with the active hydrogen attached to sulphur, as in thiols, are particularly reactive, yielding an RS (thiyl) radical

$$\phi_2 CO(^3n, \pi^*) + RSH \rightarrow \phi_2 COH + RS \qquad (8.62)$$

Because the thiyl radical is electron-poor, it is a good initiator for use with electron-rich monomers such as vinyl ethers (rather than with the electron-poor acrylates). The reactions are the basis of the *thiol-ene* process. Ketone–amine systems are widely used for photoinitiation. The ketone triplet and the amine form an exciplex (probably of a change-transfer nature) that splits into radicals by proton transfer

$$\phi_2 CO^*(T_1) + RR'NCH_2R'' \rightarrow [\phi_2 CO \overset{\delta-}{R}R'\overset{\delta+}{N}CH_2R''] \qquad (8.63)$$

$$\downarrow$$

$$\phi_2 COH + RR'NCHR$$

The tertiary amino groups in Michler's ketone, 4,4'-bis(dimethylamino)benzophenone, are specially active, but because the compound is a potential carcinogen, it has been replaced in many formulations by ethyl 4-dimethylaminobenzoate. Amines are useful co-initiators for use with ketones such as thioxanthone and its derivatives that have desirably high absorption coefficients but that also possess low hydrogen abstraction reactivity because the lowest triplets are (π, π^*) in character.

There are inherent problems associated with radical polymerization of surface coatings. Oxygen inhibits radical polymerization, an effect aggravated by the high surface-to-volume ratio in thin films, and oxygen may also quench the excited triplets of the initiator molecule (although both amine and thiol co-initiators afford some protection). Further, polymerization of double bonds involves physical shrinkage, which may affect adhesion to the substrate. Anionic polymerization is even more sensitive than radical polymerization to oxygen inhibition, and is not suitable for use in surface coatings. Cationic polymerization is much more promising. If nucleophilic species other than the monomer can be eliminated, a 'living' polymer results, in which propagation continues long after irradiation ceases, until in principle every functional group has been eliminated. Cationic polymerization is not restricted to olefinic monomers, but can also operate with strained ring systems such as cycloaliphatic and other epoxides. Little shrinkage occurs on ring opening, and slight expansion may even be possible with some monomers. Oxygen apparently does not inhibit cationic polymerization, although a very serious problem is the ease with which the propagation can be terminated by traces of nucleophilic impurity.

Photoinitiation of cationic polymerization can be achieved via charge-complex formation, using mixtures of aromatic diazonium salts and non-nucleophilic anions such as PF_6^-. Because the diazonium salts generate N_2 on photolysis, they tend to have been replaced by aromatic iodonium and sulphonium salts. One proposed initiation mechanism can be illustrated for a diaryliodonium salt such as diphenyliodonium hexafluorophosphate $((C_6H_5)_2I^+PF_6^-)$

$$Ar_2I^+X^- + h\nu \rightarrow [Ar_2I^+X^-]^*$$

$$\rightarrow ArI^+ + Ar + X^- \qquad (8.64)$$

$$ArI^+ \xrightarrow{SH} ArI^+H \rightarrow ArI + H^+ \qquad (8.65)$$

where SH represents a monomer or solvent molecule. A Brönsted acid is almost certainly primarily responsible for the initiation of polymerization, although some details remain in dispute. Although the available iodonium salts do not themselves absorb strongly in the visible or near-ultraviolet wavelengths, sensitivity in these regions can be obtained by use of photosensitizers that work by an electron transfer mechanism.

While the potential attractions of cationic photopolymerization are apparent, various commercial and technical drawbacks have limited widespread utilization of the technique. Future developments, however, seem likely to make cationic polymerization competitive with free-radical curing.

8.8.3 *Photodegradation and photostabilization*

Our brief study of the photochemistry of polymers ends with two topics related to the durability of polymers in the outdoor environment. Most organic polymers undergo chemical change, or *photodegradation*, when exposed to visible or ultraviolet radiation, especially if atmospheric oxygen is present, and as a result the mechanical properties of the bulk polymer deteriorate. Durability is essential in some contexts, such as in the building or automotive industries, so that it is desirable to extend the useful lifetime by *photostabilization* of the material. On the other hand, there is also environmental concern with the persistence of agricultural plastics and of plastic packaging materials after disposal. Polymers may therefore be made deliberately light-sensitive; use of *photodegradable plastics* may allow articles such as plastic cups to be short-lived, exposure to light reducing them to a fine powder, and thus "naturally" disposing of abandoned waste.

Most polymers do not absorb radiation strongly for $\lambda > 285$ nm, so that the sensitivity to photodegradation of hydrocarbon polymers such as polyethylene and polypropylene is, at first sight, rather surprising. Impurities, such as hydroperoxides, are formed during the high-temperature operations that convert raw polymer to the manufactured artefact. It seems that it is these impurities that lead to photodegradation, together possibly with residues of catalyst from the polymerization process. Photolysis of hydroperoxides, which absorb weakly up to $\lambda = 350$ nm, leads to the formation of carbonyl compounds that absorb much more strongly. The presence of oxygen seems essential for the photodegradation of many hydrocarbon polymers, so that the degradative process is a *photo-oxidation*. Low-temperature oxidation of hydrocarbons can proceed via a radical-chain mechanism referred to as *autoxidation*. The essential chain propagation steps can be represented by the processes

$$R + O_2 \rightarrow ROO \qquad (8.66)$$

$$ROO + RH \rightarrow ROOH + R \qquad (8.67)$$

propagated by the radical R, and in which hydrocarbon RH is oxidized to a hydroperoxide, ROOH. In the degradation of polymers, the chain is initiated by radicals, R, formed photolytically from the hydroperoxides left by the manufacturing process

$$ROOH + h\nu \rightarrow RO + OH \qquad (8.68)$$

$$OH + RH \rightarrow R + H_2O \qquad (8.69)$$

and the subsequent autoxidation leads to the catalytic production of further photochemically (and thermally) labile hydroperoxides. The type of reaction leading to the breaking of polymer chains can be exemplified by one of several routes possible for the reactions of the RO radicals from process (8.68) with polypropylene as the hydrocarbon polymer

$$\text{CH}_3-\underset{\overset{\displaystyle|}{\text{CH}_2}}{\overset{\overset{\displaystyle\text{CH}_2}{|}}{\text{C}}}-\text{O}^{\cdot} \longrightarrow \underset{\overset{\displaystyle|}{\text{CH}_3}}{\overset{\overset{\displaystyle\text{CH}_2}{|}}{\text{C}}}=\text{O} \;+\; \dot{\text{C}}\text{H}_2 \qquad (8.70)$$

ketone   radical

A polymer chain has been broken, with consequent reduction of relative molecular mass and modification of mechanical properties. In addition, one of the products is a free radical which can act directly as R in further oxidation cycles, and the other is a carbonyl compound which is readily photolysed to yield additional radicals. Unsaturated polymers are oxidized easily, probably because of stabilization of the intermediate radicals. Natural rubber (polyisoprene) is normally heavily contaminated with peroxidic impurities on account of the multiple double bonds, and is thus particularly susceptible to photodegradation.

Although photodegradation is undoubtedly more complex than the simplified version just presented, it is nevertheless clear that normal degradation is a light-initiated autoxidation process. Photostabilization of polymers must thus aim at reducing the rates of initiation or propagation of the chains, or possibly at increasing the rate of their termination. Reduction of residual impurities in the polymer would decrease the rate of initiation, and protection from oxygen would decrease the rate of propagation, but these two methods are rarely practicable.

An alternative way of reducing the rate of initiation is to prevent the absorption of light. Highly absorbent materials such as carbon-black are often used, and they confine photodegradation to the surface of the polymer. Reflecting substances such as the white oxides of zinc or titanium are used similarly. In all of these cases, the incorporation of the particulate substances may adversely affect the mechanical properties of the polymer; they may also initiate undesired photochemical processes, and they certainly limit the choice of pigmentation in the final product. Another strategy is to incorporate a soluble screen that absorbs strongly in the actinic ultraviolet regions, but that does not impart an objectionable visible coloration. Quenchers may be used that prevent the relatively long-lived triplets of carbonyl compounds from participating in the secondary photoinitiation steps. Ortho-hydroxybenzophenones make up one very useful class of stabilizer that operates by both screening and quenching. In addition, the hydroxybenzophenones seem able to react chemically with hydroperoxides, thus preventing the acceleration of autoxidation. Well-known radical scavengers such as phenols, hydroquinones, and thiols can retard photodegradation by interfering with the propagation steps.

Deliberate incorporation of photoactive groups into a polymer can render it easily photodegradable, and thus confer environmentally desirable qualities. For example, copolymerization of ketonic species with hydrocarbons yields light-sensitive polymers. Photodegradation of the resulting material does not seem to involve radical-mediated photo-oxidation, but rather Norrish type II scission of the polymer chain

$$—CH_2CHCH_2CH_2 \xrightarrow{\;h\nu\;} CH_2CH_2 + CH_2 = CH— \tag{8.71}$$
$$\begin{array}{ccc} | & & | \\ COR & & COR \end{array}$$

Ideally, disposable agricultural film or packaging should undergo a sharp and controllable degradation initiated by exposure to ultraviolet radiation. Iron (III) dithiocarbamates are potentially interesting photodegradants in this context, since at high concentrations they act as stabilizers, but at low concentrations, when they are nearly depleted, the liberated ionic iron promotes photo-oxidation. Mixtures of iron and nickel dithiocarbamate are used commercially in order to provide accurate control of both the induction period and the post-induction rate of oxidation and polymer destruction.

8.9 Solar energy storage

8.9.1 *Photochemical energy storage*

Solar energy is the ultimate source of the fossil fuels used by Man. Photosynthesis fixes about 2×10^{14} kg of carbon annually, or around ten times our present energy needs. Nevertheless, plants are not directly very useful energy sources, except perhaps in the case of wood. Rather, it is the decomposition and transformation over geological time spans that yields gas, oil, and coal. The recognition that the Earth's reserves of these fuels is not unlimited, coupled with the various oil crises of recent decades, has stimulated photochemists to see how solar radiation might be used artificially. Several aspects of such endeavours can be recognized, including (i) the storage of energy for later release, as heat, light, or electricity; (ii) the immediate production of electrical power; and (iii) the formation photochemically or photoelectrochemically of products that would otherwise require consumption of power from other sources (such as production of Cl_2 from the Cl ion). In this section, we outline the principles underlying some of the purely photochemical methods being considered in the goal of solar energy conversion, and defer discussion of electrochemical methods until the following section. The division is somewhat artificial, since, as will appear, the products and the chemistry involved are often closely related in the two cases.

Certain features are common to all forms of photochemical solar energy utilization, and can conveniently be considered in their general forms here. The 'photochemical' processes all naturally involve the generation of electronically excited states, and the objectives are to use those excited states to form products (including electrons and ions) that are energy-rich with respect to the reactants. From a thermodynamic viewpoint, energy beyond the energetic threshold for formation of the required products is 'wasted', since, in condensed-phase systems at least, excess energy will be degraded to heat in the absorbing medium. There is, however, another side to the question of excess energy. Whenever high-energy products are formed by an 'uphill' process, they are thermodynamically unstable with respect to the starting materials. The products can be harvested only if there are kinetic constraints to the reverse reaction. Two obvious constraints are that the reverse (exothermic) reaction must have an appreciable activation barrier, which implies that the forward reaction also has a barrier greater than the endothermicity of reaction, or that the products must move apart fast and far enough not to re-encounter each other before they can be utilized, a situation again implying an input of energy greater than the thermodynamic threshold value. Purely homogeneous systems are at a serious disadvantage in this respect, in that the newly formed products are likely to be trapped within a solvent cage. In heterogeneous systems (or microheterogeneous ones such as micelles), the possibility arises that the products can be kept far enough apart to prevent the back reaction. We shall, however, start by examining photochemistry in homogeneous solutions because it provides a guide to what might occur in the more complex cases.

A further critical factor in solar energy conversion is that the absorption spectrum must match the spectrum of the Sun's radiation incident on the Earth's surface. Optimum threshold absorption limits can be shown to lie in the range 1100–700 nm, with theoretical limiting thermodynamic efficiencies of around 30% for single-photon processes. It is worth pointing out that, although 700 nm is close to the threshold of absorption in green plants, the average efficiency of natural photosynthesis is nearly two orders of magnitude smaller than the thermodynamic limit.

It is apparent that any useful method of photochemical energy conversion must either use cheap and abundant raw materials, perhaps to generate storeable fuels, or must be capable of operating cyclically with regeneration of the energy-carrying material. We shall consider examples of each of these possibilities.

One type of energy storage that has been proposed uses photoisomerization of organic molecules from a low-energy to a higher-energy structure. One widely studied system is the photoisomerization of norbornadiene (A) to quadricyclane (B)

$$\xrightarrow{h\nu}$$ (8.72)

A   B
Norbornadiene   Quadricyclane

The enthalpy of isomerization from A to B is about $110 \, \text{kJ mol}^{-1}$, but the activation barrier is much higher, so that there is also a barrier to the reverse reaction, and B is kinetically quite stable with respect to A. Energy can therefore be released at will in a catalysed reversion process. One particular drawback in this system is that norbornadiene absorbs only in the ultraviolet region at wavelengths shorter than those available from the Sun, although use of substituted norbornadienes together with suitable triplet photosensitizers can shift the active region to visible wavelengths. Other reasonable isomerization systems exist, including the indigoid dyes, that absorb visible wavelengths directly. All these organic photoreactions suffer from the disadvantage, however, that there are side reactions in the forward and catalysed reverse steps that ultimately cause depletion of the reactants. Practical development of isomerization storage of solar energy thus seems to founder on the high costs of the complex organic compounds that can only undergo a limited number of cycles.

Of possible consumable raw materials, water, carbon dioxide, and nitrogen must be amongst the cheapest and most abundant. Possible reactions include the 'natural' photosynthetic process itself

$$CO_2 + H_2O \rightarrow (CH_2O) + O_2 \quad (\Delta G^{\ominus} = 496 \, \text{kJ mol}^{-1}) \quad (8.73)$$

and the two analogous reactions

$$\tfrac{2}{3}N_2 + 2H_2O \rightarrow \tfrac{4}{3}NH_3 + O_2 \quad (\Delta G^{\ominus} = 452 \, \text{kJ mol}^{-1}) \quad (8.74)$$

$$2H_2O \rightarrow 2H_2 + O_2 \, (\Delta G^{\ominus} = 472 \, \text{kJ mol}^{-1}) \quad (8.75)$$

where the free energy change, $\Delta G^{\ominus}$, indicates the amount of energy available in the products for doing useful work. One can envisage conversion of carbohydrate to alcohols for use as liquid fuels, following current practice with natural carbohydrates, and the fixation of atmospheric nitrogen to ammonia could replace the energy-greedy methods now used to make, for example, fertilizers. The real prize, however, is the photochemical fission of water, because hydrogen is a fuel that can provide pollution-free combustion, as well as being a feedstock for electricity-generating fuel cells.

Direct photolysis of water is not a candidate for solar energy conversion, since water does not absorb in the visible spectral region; fission to form $H + OH$ as radical fragments has an energy threshold around $\lambda = 240$ nm, and even at that wavelength absorption is weak. An ionic redox mechanism, on the other hand, requires the transfer of four electrons so that each transfer demands a free energy of $472/4 = 118 \, \text{kJ mol}^{-1}$, corresponding to a wavelength per photon of roughly 1000 nm in the near-infrared (or, in electrode potential terms, about $118\,000/96\,500 = 1.22$ V). Such a multiphoton redox fission of water thus seems promising, and the question is how to make it occur.

The reactions involved would formally be

$$4H^+ + 4e \rightarrow 2H_2 \tag{8.76}$$

and

$$2H_2O \rightarrow O_2 + 4e + 4H^+ \tag{8.44}$$

The redox potentials for these two reactions are -0.41 V and $+0.81$ V, respectively, at pH $= 7$; the total, of course, is the 1.22 V already calculated. The presence of a redox couple of potential less than -0.41 V will cause the evolution of H_2, while the presence of a couple with a potential greater than $+0.81$ V will result in the evolution of O_2.

Absorption of light by one or other of the partners in a redox pair alters the redox potential because of the excitation of electrons. Excitation of the reducing partner makes electrons more readily accessible, and thus reduces the redox potential by an amount equivalent to the electronic excitation energy. Conversely, electronic excitation of the oxidizing partner increases the redox potential, because the electron hole left by the promoted electron makes the excited molecule a better electron acceptor. One might thus

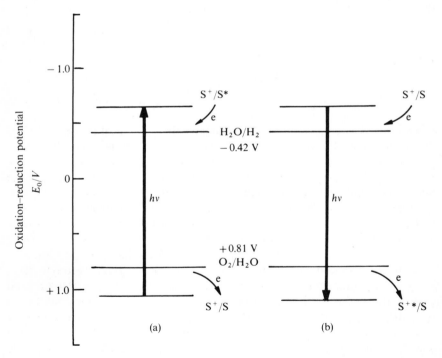

Fig. 8.16 The simultaneous photo-oxidation and photoreduction of water: redox schemes in which the light is absorbed (a) by the reductant and (b) by the oxidant partner of the couple S^+/S.

imagine a redox pair that could oxidize water when unexcited, but that could reduce water when the reducing partner absorbed radiation. In an exactly similar way, a normally reducing pair might oxidize water on excitation of the oxidizing partner. These ideas are represented in Fig. 8.16.

Unfortunately, real systems have yet to be found that operate quite so simply, and many ingenious indirect routes have been devised in order to overcome the various kinetic and energetic limitations. It has proved generally difficult to run both oxidation and reduction processes, reactions (8.44) and (8.76), with the same redox pair. Instead, many experiments have concentrated on one or other components, especially the less demanding reducing step. In that case, an external source of electrons must be found if the absorbing species is not to be consumed (i.e. if the redox system is to work catalytically). That source may be a chemical species that acts as a 'sacrificial donor'. For the analogous photo-oxidation process, a sacrificial acceptor is needed to remove electrons. Colloidal platinum can be used to trap electrons from one-electron reduced species, and so couple together the photochemistry and water reduction. Successful systems have also used an intermediate acceptor, or 'electron relay', between the photosensitizer and the platinum-/water reactant. Using the symbols S for sensitizer, D for sacrificial donor, and A for intermediate acceptor, the general scheme is thus

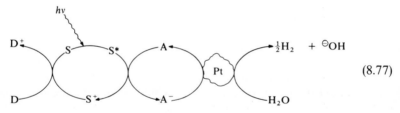

$$(8.77)$$

These general remarks may be given more substance by illustrating them with a real and successful photoreduction system. A redox couple much used in water splitting experiments involves a tris(bipyridyl) complex of ruthenium, represented by $Ru^{3+}(bipy)_3/Ru^{2+}(bipy)_3$, and which we shall further simplify to R^{3+}/R^{2+}. The couple has a redox potential of 1.27 V when unexcited, but the R^{2+} absorbs strongly in the visible region with an excitation energy of about 2 eV, so that the redox potential becomes $(1.27-2) = -0.83$ V. Two of the attractions of the ruthenium complex pair are the strong absorption and the relatively long lifetime of the R^{2+*} excited state that permits reaction with other partners. An effective intermediate acceptor is the compound methyl viologen (paraquat), represented as MV^{2+} changing to MV^+. The sequence of events is thus

$$4R^{2+} + 4h\nu \rightarrow 4R^{2+*} \qquad (8.78)$$

$$4R^{2+*} + 4MV^{2+} \rightarrow 4R^{3+} + 4MV^+ \qquad (8.79)$$

$$4MV^+ + 4H^+ \xrightarrow{\text{Pt}} 4MV^{2+} + 2H_2 \qquad (8.80)$$

Finally, the reduced form of the ruthenium couple is regenerated by a sacrificial donor also present, such as ethylenediaminetetracetic acid (EDTA), which also inhibits the reverse of reaction (8.79) by removing the oxidized ion

$$4R^{3+} + 4EDTA \rightarrow 4R^{2+} + 4EDTA^+ \qquad (8.81)$$

The net result of this set of reactions is thus the photosensitized reduction of water to yield molecular hydrogen, as desired, neither the ruthenium sensitizer nor the MV^{2+} being consumed.

The concepts behind the photochemical oxidation of water are similar to those already explained, although greater problems are found in the practical implementation because of the need to transfer four electrons rather than two. Sacrificial acceptors are needed in this case, and persulphates have been found effective. The key in oxygen production is the use of a "super colloid" of RuO_2 as catalyst to effect the O_2 liberation, just as Pt is used in the reduction system. The oxygen production scheme is thus

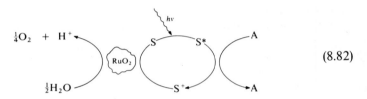

$$(8.82)$$

Some success has been reported with metalloporphyrins as sensitizers in place of the ruthenium complex, which would be an advantage in commercial exploitation of the process. Water-soluble zinc porphyrin looks particularly promising, giving quantum yields for O_2 production approaching 0.5. An even greater challenge than just producing oxygen is to link the reduction and oxidation systems, so that no sacrificial compounds are required. It will be recalled that the natural photosynthetic process (Section 8.3) achieves an equivalent linkage by using a common redox system operating between the two pigment systems. Attempts to mimic these processes in the laboratory have generally failed because of the need to provide kinetic competition between the desired forward reaction and the competing back reactions. Amongst the proposals made to overcome these difficulties are several involving organized structures such as micelles made up of hundreds of surfactant molecules, and the separation of the two reactions in space, for example by the use of membranes that permit transport of electrons and protons but not of larger species.

In view of the difficulties experienced with the water system, other materials such as dilute ethanol from low-grade fermentation look increasingly attractive. Good yields of H_2 and CH_3CHO have been obtained with benzophenone as sensitizer, and with a highly active Pt colloid as catalyst. Although the absorption of benzophenone is a poor match for the solar spectrum, since it absorbs mainly in the ultraviolet, the excited sensitizer reacts so rapidly with ethanol that the need for an electron relay such as MV^{2+} is obviated. The reaction is believed to involve the formation of $(C_6H_5)_2C(OH)$ and CH_2CH_2OH radicals in the interaction between excited benzophenone and ethanol, followed by the reaction of the two radicals on the Pt surface to yield H_2 and CH_3CHO and to regenerate the benzophenone. The quantum efficiency for ethanol-splitting approaches unity, so that the overall solar conversion efficiency is roughly the fractional absorption ($\sim 5\%$) of the incident solar energy. Most natural photosynthetic processes have an efficiency lower than that of this artificial system.

8.9.2 *Photoelectrochemical energy storage*

Solar energy storage using photoelectrochemical processes can, in principle, be achieved in several ways. First, irradiation of chemical species in homogeneous solution might produce electrical energy in *photogalvanic cells*. Secondly, irradiation of an electrode immersed in a solution can yield electrical power because a *photovoltaic junction* is formed at the interface. Next, such a cell might be made to yield chemical products by completing the external circuit by joining the two electrodes to perform *photoelectrolysis*, analogous to the electrolysis that occurs when an external electrical power source is used. Finally, the individual electrodes might be dispensed with, and be replaced by photosensitive particles in suspension that combine on each particle the characteristics of anode and cathode, so that electrolysis (e.g. of water) can occur 'microheterogeneously' throughout the solution.

Of these possibilities, the simplest to describe is the photogalvanic cell, although it is also the least promising from a practical aspect. The cell consists of two metal electrodes and an electrolyte containing a dye and a redox couple. Let us take as an example a cell with platinum electrodes, the dye tris(2,2′-bipyridyl)ruthenium(II) (given the symbol R^{2+} as in the previous section), and with Fe^{3+} as an electron acceptor. In the dark, the equilibrium

$$R^{2+} + Fe^{3+} \rightleftharpoons R^{3+} + Fe^{2+} \qquad (8.83)$$

is set up. Since the redox potential of the ruthenium couple is 1.27 V (see Section 8.9.1), and that of Fe^{3+}/Fe^{2+} is 0.7 V, the equilibrium is shifted almost entirely to the left. There is, of course, no potential difference between the two equal electrodes. On illumination R^{2+*} is formed, and an electron transfer to Fe^{3+} can now displace the two redox systems from equilibrium. There is, however, a back reaction that leads ultimately to the establishment of a photostationary state. Although the potentials at the individual electrodes may nominally alter, no potential difference can be discerned if the illumination is even. If, on the other hand, the incident light is absorbed nearer one electrode than the other, then a photovoltage may be observed, because something rather like a concentration cell has been set up. The occurrence and sign of the photovoltage depend on which of the two redox couples is more reversible at the electrodes, i.e. on the electrode kinetics. Efficiencies of energy conversion are low ($<1\%$) in such devices, primarily because of disadvantageous electron transfer kinetics. The high-energy redox products created by irradiation try to revert to the starting material by back electron transfer, rather than in the desired fashion of passing electrons through the load of the external circuit.

Photoredox reactions at the interface between semiconductors and liquid electrolytes give much more impressive results than the homogeneous proc-esses. These processes are the chemical equivalents of the events that take place in solid-state junction photovoltaic devices. Silicon 'solar cells' are now well established as power sources, their applications ranging from driving the instrumentation on space probes, to running signalling equipment on remote railway tracks, to powering pocket calculators. The goal in developing chemical semiconductor solar energy conversion must be to outperform, in terms of yield or manufacturing cost, the solid-state solar cells. In order to explain the operation of semiconductor photoelectrochemical cells, as well as / to compare chemical and physical devices, we should remind ourselves briefly of the properties of semiconductor junctions and of how an electrical poten-tial can be set up by shining light on such a junction.

Silicon is commonly used for solar cells, and we shall initially illustrate our explanation with respect to it, although other semiconductor species such as GaAs, CdS, TiO_2, etc., will appear later in our discussion. In solids such as silicon, the isolated atomic or molecular orbitals that are familiar in gaseous and liquid species coalesce as a result of lattice interactions to become 'bands' separated in energy by regions where there are no energy levels at all. The lowest energy band is the *valence band*, which is filled with electrons. The next higher band is the *conduction band*, and electrons can be promoted to it by absorption of radiation, amongst other means. Promotion of the electron leaves behind a positive 'hole' in the valence band. The separation between the highest energy in the valence band and the lowest energy in the conduc-tion band is the *band gap*, and corresponds to the threshold wavelength at which absorption will promote an electron. Substitution of some of the Group IV Si atoms with atoms of a Group V element such as P leads to the formation of *n-type* doped material, whose electrical properties are modified by the dopant atoms. The additional electron of the pentavalent dopant cannot be accommodated in the filled valence band, and thus becomes an itinerant electron in the conduction band. Similarly, doping Si with a Group III material, such as B or Al, leads to the formation of *p-type* material, in which there are vacant electron sites, or *holes*, in the valence band. Holes are also itinerant in the sense that they can propagate by capturing electrons, the sites of which now become new holes. The electrons have, loosely speaking, an average energy, which is just below the conduction band in the electron-rich n-type material and just above the valence band in the electron-deficient p-type silicon. The average energy is, in reality, the chemical potential of the electrons, a quantity familiar from thermodynamics. Solid-state physicists call this chemical potential the *Fermi level* of the electrons. In a liquid electrolyte the quantity would be called the redox potential. Figure 8.17(a) indicates the energy relations between the valence band, conduction band, and Fermi level for n- and p-type semiconductors.

Making a junction between n- and p-type materials leads to the novel properties that make solid-state electronics possible. Since the chemical potential of the electrons must be identical on each side of the junction, it follows that the levels of both the conduction and the valence bands of the p-type material must be displaced to higher energies, as illustrated by Fig. 8.17(b), and the two bands bend upwards over the junction region as the material becomes progressively more p-type. Electrons migrating in the conduction band from p- to n-sides do so readily since they are moving 'downhill' to a region of lower energy. Likewise, holes move to lower hole energy (reverse sense to electron energy) in migrating from n- to p-sides. Migration of holes or electrons in the opposite directions is impeded by the potential barrier imposed by the displacement of the bands. We note that the separation of charges implied by this picture means that, at equilibrium in the dark, there is a potential across the junction, the n-material being negative with respect to the p-material, and that the junction region itself will be depleted of charge carriers.

When illuminated by radiation, a p–n junction functions as a solar cell that can promote electrons from the valence to the conduction bands. If the bulk n- and p-type silicon are connected through an external load, electrons will flow as long as light is absorbed and useful work can be done. An idealized diagram of a photovoltaic cell is given in Fig. 8.18(a), and the mechanism of the photovoltaic effect is shown in terms of the semiconductor energy levels in Fig. 8.18(b). Optical excitation of an electron to the conduction band creates an electron–hole pair. Both electron and hole possess potential energy in the junction region (by virtue of the electrostatic field present) and can lose some of that energy, the electron by moving towards the n-type material and the hole by moving towards the p-type. Thus the n-type silicon becomes the negative pole and the p-type silicon the positive pole of the solar cell. The electrical potential energy, U_p, created by the photons now adds to the electron energy so that the Fermi levels in the n- and p-type materials become separated by U_p. A very important point to note is that the charge-pair, once created, experiences forces that tend to separate the components rather than to permit their loss by recombination.

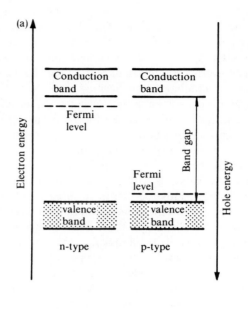

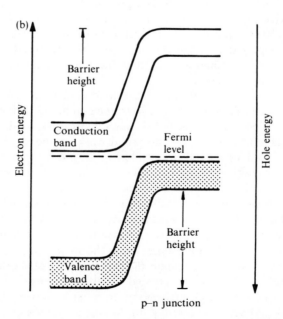

Fig. 8.17 Energy relations between the valence band, conduction band, and Fermi level: (a) in isolated n- and p-type semiconductors; (b) where a p–n junction is made.

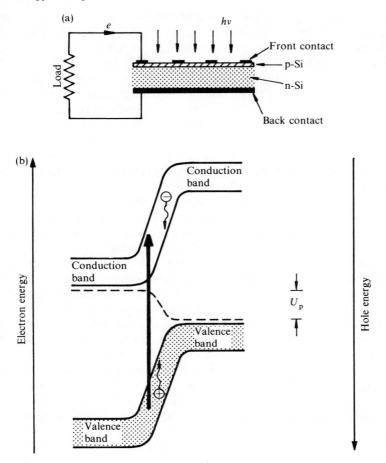

Fig. 8.18 (a) Idealized diagram of a semiconductor photovoltaic cell. (b) Mechanism of the photovoltaic effect in terms of the semiconductor energy levels.

We are now in a position to understand how the semiconductor–liquid junction operates. When a semiconducting electrode is immersed in a solution containing a redox couple, the chemical potentials of both electrode and solution must be identical if no external force is applied. Once again, then, the bands in the semiconductor bend so as to equalize the Fermi level and the redox potential. The direction of bending depends on the particular system, but for n- and p-type materials it is usually in the direction indicated in Fig. 8.19(a) and (c). Illumination of the electrode surface with light can bring about promotion of electrons from valence to conduction bands, and the field gradients at the junction will, as in the purely solid-state case, bring about separation of the newly created electrons and holes. For upward curvature, as in Fig. 8.19(a), electrons move into the bulk of the semiconductor, while holes leave the surface to oxidize the redox couple. If, then, an external circuit is made to a counter-electrode immersed in the solution, then electrons will flow from the semiconductor to the counter-electrode (to reduce ions in solution there), so that the semiconductor electrode has become a *photoanode*, as in Fig. 8.19(b). As a result of the electrochemical potential, U_p, resulting from the photovoltaic effect, Fermi and redox potentials become separated by U_p. Figure 8.19(d) shows the analogous energy diagram for absorption by the p-type material, from which electrons enter the solution to reduce the redox couple, and which is therefore a photocathode.

One major problem with the semiconductor electrolyte cells concerns corrosion of the semiconductor by the holes or electrons. For this reason, bare silicon itself is not a suitable material for practical electrochemical cells because it can readily be oxidized to SiO_2, although 'derivatization' with, for example, ferrocene stabilizes the electrode. As another example, a cell with an n-CdS electrode will not operate for very long during irradiation, because the photogenerated holes at the surface cause decomposition of the CdS to $Cd^{2+} + S$, and the surface soon becomes covered with insulating sulphur. Some protection can be afforded by choice of a suitable redox couple. In the case of CdS, high concentrations of polysulphide or polyselenide ions (S_n^{2-}, Se_n^{2-}) can stabilize the photoanode, because the solution reductant, rather than the lattice S^{2-}, is oxidized, and the S_n^{2-} species are reduced at the counter-electrode to complete a circuit involving no net chemical change. Layered structures of MoS_2, $MoSe_2$, and WSe_2 are proving to have better stability than more conventional semiconductors. Stable and efficient cells have been made that use $MoSe_2$ photoanodes in iodide–triiodide redox electrolytes.

The diagrams of Fig. 8.19 should make it clear that the maximum output voltage, for no load and high irradiance, could reach $E_f - E_v$ for a photoanode and $E_f - E_c$ for a photocathode, where E_f, E_v, and E_c are the potentials of the Fermi level, valence band, and conduction band, respectively. Further, E_f has a maximum value approaching E_c for a highly doped n-type semiconductor, and approaching E_v for a highly doped p-type one. Thus, the absolute maximum voltage that could be envisaged would be equal to $E_c - E_v$, or the band gap of the semiconductor material. This limit does, however, presuppose that E_{redox} is nearly as low as E_v in the n-type system (or as high as E_c in the p-type system). Band bending depends on the difference between the band and E_{redox} in the free semiconductor and electrolyte, so that these limits cannot be approached without losing the desired separation of the newly formed electron–hole pairs and the consequent inhibition of recombination. Bending of about 0.3 V seems to be sufficient to obtain the maximum photocurrent. In terms of solar energy conversion, the most efficient

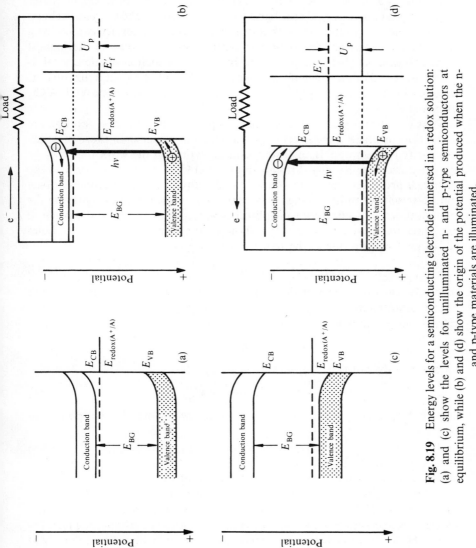

Fig. 8.19 Energy levels for a semiconducting electrode immersed in a redox solution: (a) and (c) show the levels for unilluminated n- and p-type semiconductors at equilibrium, while (b) and (d) show the origin of the potential produced when the n- and p-type materials are illuminated.

processes will involve semiconductor band gaps corresponding to the solar spectral range of 1100–700 nm already discussed in Section 8.9.1; these wavelengths correspond in electron energy units to about 1.2–1.8 eV. As the band gap of a semiconductor drops below 1 eV, the photocurrent increases but the voltage is low, while for band gaps above 1.6 eV, the current declines. There is, incidentally, evidence that dyes attached to the surface can increase the current output of wide band-gap semiconductors. Some of the highest solar efficiencies (11–14%) are shown by materials such as single-crystal n-GaAs in an alkaline selenide electrolyte (Se_2^{2-}/Se^{2-}) for stabilization as described in the last paragraph, and by surface-treated p-InP in a VCl_3/VCl_2 electrolyte.

Possibly of more interest from the viewpoint of solar energy storage are those cells in which there is net chemical reaction rather than those in which electricity is generated. If, instead of concentrating on the work that may be done by electrons passing through an external load, we consider instead the chemical processes occurring at anode and cathode, it soon becomes apparent that useful chemical change may arise, even to the extent of achieving the goal outlined in Section 8.9.1 of splitting water into hydrogen and oxygen. If the semiconductor electrode is the photoanode, then the liberation of holes (h^+) there oxidizes one of the redox partners

$$A_{red} + h^+ \rightarrow A_{ox} \qquad (8.84)$$

where A refers to species at the anode. Transference of the electron through the external circuit to the counter-electrode (typically carbon or a metal) means that some oxidized component, C_{ox}, can be reduced to C_{red}

$$C_{ox} + e \rightarrow C_{red} \qquad (8.85)$$

so that the overall reaction, promoted by light, is

$$A_{red} + C_{ox} \rightarrow A_{ox} + C_{red} \qquad (8.86)$$

If both A_{red} and C_{ox} were water, then reaction (8.86) would be the splitting of water

$$H_2O + H_2O \rightarrow O_2 + 2H_2 \qquad (8.87)$$

which was proposed in Section 8.9.1 as one of the goals of solar energy conversion. Since the potential energy difference between H_2O/H_2 and O_2/H_2O is 1.22 V, there is a good match with the solar spectrum, and efficiencies in excess of 20% might be possible. Photoelectrosynthetic production of H_2 and O_2 in reaction (8.87) has been achieved with cells using n-TiO_2 or, more particularly, n-$SrTiO_3$ photoanodes. Unfortunately, $SrTiO_3$ has a band gap of about 3 eV, so that it can respond only to ultraviolet radiation. The efficiency of converting solar to chemical energy is therefore only about 1%, although the ultraviolet light itself is converted with about 25%

efficiency to storable energy. Several other comparable reactions have been discovered, including the reduction of CO_2 to HCHO and CH_3OH with p-GaP as semiconductor. In most cases, an electrical bias has to be provided, so that these reactions are *photoassisted electrolyses*. Yet another category of process is that in which a downhill reaction (i.e. $\Delta G^\ominus$ negative) is driven photoelectrochemically, the photon energy being used, in these cases, to overcome kinetic activation barriers rather than being stored. For example, the reaction

$$2CH_3COOH \rightarrow C_2H_6 + 2CO_2 + H_2 \qquad (8.88)$$

has, perhaps surprisingly, a negative $\Delta G^\ominus$. The photoelectrochemical decomposition has been carried out using a n-TiO_2 photoanode.

An extension of the ideas of photoelectrochemistry involving interfaces is the use of semiconductor particles. Powders of TiO_2 with platinum dispersed on the surface have been particularly successful. Each particle can be thought of as a closed-circuit photoelectrochemical cell, the semiconductor- and counter-electrodes being in contact. The basic principles outlined above thus still apply, even though there is no external circuit. Although the distance between anode and cathode is much less than in conventional electrochemical cells, the products of charge transfer reactions are still kept apart in a way that is not possible with homogeneous processes when both partners are formed within the same solvent cage. Several heterogeneous photosynthetic and photocatalytic processes that use particulate semiconductors have been described, including the production of CH_3OH from CO_2, RH from RCOOH, and NH_3 from N_2. In some cases, the semiconductor powders alone appear to act as photocatalysts, the metal islands being unnecessary. Yields are rather low, usually because of kinetic limitations and because large-band-gap materials must be used that do not efficiently utilize the solar spectrum. It may be that the strategy of natural photosynthesis is still needed, and that losses must be made good by using *two* lower-energy photons to transfer one electron.

One question that needs to be answered in relation to photoelectrochemical exploitation of solar energy is whether any advantage accrues over using straightforward solid-state junction devices to produce electricity, either for immediate consumption or to electrolyse water or other materials to provide fuels for later use. The efficiency of the best solid-state cells still seems somewhat better, at nearly 20%, than that of the best electrochemical cells, and the solid-state cells do not suffer from the lack of stability that limits the performance and lifetime of practical electrochemical cells. So what are the perceived advantages of the photoelectrochemical route to solar energy storage? One is that the junction is produced very easily just by immersing the semiconductor in the redox solution. Another is that light is absorbed directly at the interface, so that the charge carriers produced by radiation need only move a short distance before they reach their reaction partners, whereas in the n–p cell discussed earlier, light has to penetrate the thin p-layer, and many photogenerated carriers have to diffuse to the junction depletion zone where separation occurs. It has been thought that high-grade (and high-cost) single-crystal material is needed in such circumstances in order to prevent recombination of charge carriers at, for instance, grain boundaries, with a concomitant loss of efficiency. Polycrystalline, and perhaps even sintered, materials suffice for the electrochemical cells. The advantages of electrochemistry are thus seen essentially as economic. However, at the time of writing, the whole stimulus to research in photochemical energy conversion is being challenged, because some semiconductor manufacturers, notably in Japan, have obtained high efficiency (ca. 10%) with purely solid-state devices using multilayer amorphous silicon. The devices can be made geometrically large and at an economically competitive price. If these developments continue, then solar energy conversion may really move towards electrolysis of water to store energy produced by solid-state junction devices. The search for chemical storage schemes will not have been wasted because the fundamental chemistry that has been uncovered has shown photochemists another fascinating facet of their discipline.

8.10   Photochemistry in synthesis

Photochemistry has important applications in synthesis in the chemical industry. A very few examples must suffice here to show the types of compound to which photochemical routes are suited. The main advantage of photochemical reactions in synthesis is that highly specific reactions can be brought about by light, to give products that would be difficult or impossible to form using thermal reactions. In the laboratory, photochemical methods are much used in the formation of four-membered rings by [2 + 2] inter- and intramolecular cycloaddition and by electrocyclic ring closure of conjugated dienes. Photochemical techniques also offer a route to highly strained compounds whose thermodynamic instability disfavours thermal methods. Chapter 6 will indicate the scope of photochemistry for laboratory synthesis.

   We are more concerned here with the use of photochemistry in industrial synthesis. A photochemical process must obviously be superior in yield or purity of product in order for it to be competitive with alternative thermal methods. Reactions proceeding by a chain mechanism (often with radical chain carriers), in which the initiation step is photochemical, are particularly favourable candidates for industrial use. Indeed, we have already seen this application in connection with photopolymerization (Section 8.8.2). However, a photochemical reaction may even be economic if the quantum yield is low, so long as the chemical yield is higher than that available from thermal processes. In *fine*-chemical manufacture, the use of light may represent only a small fraction of the total cost of the high-value product. Furthermore, the relatively small quantities of material involved mean that a batch process can often be used that is a scaled-up version of the laboratory method. Greater difficulties arise with the use of photochemistry in large-scale *heavy*-chemical manufacture, because the energy cost may now be a substantial element in the cost of the final product. Continuous-flow reactors are often used in large-scale processes, which poses some problems of design in photochemistry. In particular, transparent reactors or transparent lamp housings must be used, the walls of which often become contaminated by the build-up of tarry (and opaque) by-products; the absorption of radiation by the reactants may also seriously limit the size of the reactor. Against these undoubted disadvantages of photochemical synthesis must be set the greater selectivity of products and the better control over their formation. Lower thermal loads are imposed in the manufacturing process, since the reactants do not need to be heated and then cooled. Techniques to reduce the impact of the problems with photochemical reactors have been developed. They include irradiation of the surface of falling thin films of reactant, the use of laminar flows of non-miscible fluids, the one further from the reactor wall being the light-absorbing one, and the use of gas bubble-induced turbulence to improve reactant exchange. Finally, the cost of using light energy should not be overstated: in favourable cases, more than one mole of product can be obtained for the expenditure of one kilowatt-hour of electrical energy, even for reactions in which there is no chain process.

Chain reactions do play a part in one of the most important applications: the chlorination of hydrocarbons. The process is illustrated by the reactions

$$Cl_2 + h\nu \rightarrow Cl + Cl \tag{8.89}$$

$$Cl + RH \rightarrow R + HCl \tag{8.90}$$

$$R + Cl_2 \rightarrow RCl + Cl \tag{8.91}$$

Kinetic chain lengths (see Section 1.8) can be as large as 1000: i.e. the overall quantum yield is very large, so that a relatively small radiation source can be used for high outputs of chlorinated material. In alkylated aromatic systems (e.g. toluene), photochlorination allows substitution in the alkyl group to the exclusion of chlorination of the aromatic ring. With benzene itself, addition occurs, to make hexachlorocyclohexane. The γ-isomer is a valuable biologically degradable insecticide, commonly known as gammexane or lindane. Photochemical chlorination has the advantage over thermal synthesis that, at the relatively low temperatures required, the γ-isomer is formed in higher yields relative to the other isomers, which are contaminants.

Photosulphochlorination may occur, with long chains and high overall quantum yield, if sulphur dioxide is present together with chlorine and the hydrocarbon. Two new steps replace reaction (8.91)

$$R + SO_2 \rightarrow RSO_2 \qquad (8.92)$$

$$RSO_2 + Cl_2 \rightarrow RSO_2Cl + Cl \qquad (8.93)$$

Sulphonyl chlorides, the products of reaction (8.93), are intermediates in the manufacture of alkanesulphonates, which are used as surfactants and emulsifiers.

One example is often quoted of a large-scale process that is commercially successful in spite of having a quantum yield of less than unity. It is the photo-oximation of cyclohexane to yield ε-caprolactam. Caprolactam is the cyclic amide that produces, by ring opening and addition, Nylon-6, which is a Nylon variant finding high-tonnage applications

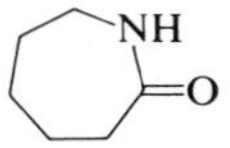

$$\longrightarrow \quad -[NH(CH_2)_5CO]_n- \qquad (8.94)$$

Caprolactam

In the manufacture of caprolactam, nitrosyl chloride (NOCl) is the photo-chemically active species used. Visible radiation will bring about fission of the weak Cl–NO bond, and the subsequent reactions of Cl and NO fragments with cyclohexane lead to the formation of an oxime and finally caprolactam.

$$NOCl + h\nu \longrightarrow NO + Cl \qquad (8.95)$$

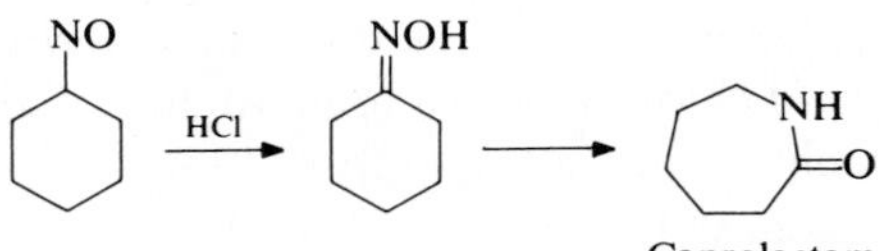

$$Cl + \longrightarrow HCl + \qquad (8.96)$$

$$+ \; NO \longrightarrow (+Cl) \qquad (8.97)$$
(or NOCl)

$$(8.98)$$

Caprolactam

Hydrogen chloride is added to the nitrosating gas, and the ratio of NOCl to HCl is critical in determining the yield of the desired product. The cost of manufacturing caprolactam photochemically from 'inert' hydrocarbon compares favourably with that of producing the material by thermal reactions from starting materials such as phenol or toluene. Nylon 12, $-[NH(CH_2)_{11}CO]-$, has important special uses for dimensionally stable articles, foodstuffs packaging, and coatings on metals. Its precursor, lauryl lactam (dodecanolactam) can also be produced by the photochemical route, this time using cyclododecane as the starting material.

Photochemical production of fine-chemicals is typified by the manufacture of vitamin D_3, a material used extensively in animal nutrition. 7-Dehydro-cholesterol, made from cholesterol, undergoes electrocyclic ring opening on ultraviolet irradiation. The resulting triene, pre-vitamin D_3, is converted to vitamin D_3 itself by a thermally-induced 1,7-hydrogen shift

7-Dehydrocholesterol   Previtamin D_3

$$(8.99)$$

Vitamin D_3

Several other physiologically active substances can be prepared photo-chemically. For example, prostaglandins, hormones having considerable chemotherapeutic value, can be obtained by a synthesis starting with the photolysis of cyclic ketones. In another area of manufacture, the stereo-isomeric rose oxides, used in perfumery, are obtained by the photo-oxidation of citronellol. Excited (singlet) oxygen (see pp. 149–151) is formed by photosensitization with a dye such as rose bengal, which transfers energy to ground-state (triplet) O_2 to conserve total spin. Hydroperoxides are formed by addition of singlet oxygen to the double bond, and subsequent reduction yields the corresponding alcohols. Allylic rearrangement in acid solution, followed by dehydration, leads to the desired product.

Citronellol   Rose oxide

$$(8.100)$$

Lasers as light sources might be expected to find wide application in industrial synthesis. However, the high-power lasers needed are not yet available commercially, and laser methods are restricted in industry to selective separation of molecules at the atomic or molecular level. Photochemical separation of isotopes (cf. p. 63) is one such use. Laser isotope separation depends on the shifts in the optical absorption spectrum resulting from isotopic substitution. The highly monochromatic radiation from a laser can be used to excite only those molecules containing a particular isotope. Some means is then provided of removing or harvesting the excited species. For example, ^{35}Cl and ^{37}Cl may be separated by exciting ICl with radiation from a dye laser and allowing the excited molecules to react with 1,2-dibromoethene. Some 1,2-dichloroethene is formed, the simplified scheme

$$ICl + h\nu \rightarrow ICl^* \tag{8.101}$$

$$2ICl^* + C_2H_2Br_2 \rightarrow C_2H_2Cl_2 + 2IBr \tag{8.102}$$

indicating the nature of the interactions. If the laser is tuned to the absorption of $I^{37}Cl$ ($\lambda = 605.4$ nm), the resulting dichloroethene shows an enrichment in ^{37}Cl of nearly 20 over the natural isotopic abundance. A major stimulus to developing methods for laser isotope separation has been the need for the ^{235}U used in nuclear reactors. One enrichment scheme employs two-step photoionization of a molecular beam of uranium, in which the first laser selectively excites neutral U atoms, and the second photoionizes the excited (but not the ground-state) uranium. The beam contains neutral molecules as well as ions, but the ions are preferentially of the 235-isotope. Electrostatic separation therefore enables the enriched sample to be collected.

8.11   Optical brighteners

Coloured fluorescent dyes and pigments find many non-scientific applications: they are used for the brilliant 'dayglo' paints, for textile dyes, and to obtain special theatrical effects. No application is so widespread, however, as the use of special fluorescent substances as *optical brighteners* or *bleaches* in the 'whiter-than-white' washing powders. The principle behind the operation of an optical bleach is that the substance should absorb in the ultraviolet and radiate in the visible region, so that the washed (white) textile apparently reflects more light than was incident upon it. A related large-scale application is in the optical bleaching of paper.

A substance that is to be suitable as an optical bleach must satisfy several stringent requirements. First, it must not absorb at all in the visible, since this would lead to coloration of the fabric, but must absorb strongly in the near-ultraviolet, where there is still some intensity available from natural or artificial light sources. Secondly, the fluorescence must lie in the short-wavelength end of the visible spectrum, as otherwise the fluorescence would give an apparent undesirable yellowing to white fabric. Thirdly, the fluorescent substance must be photochemically stable, and it must not sensitize degradation or oxidation of the fibre material. Lastly, the substance must be soluble or dispersible in the aqueous detergent solution, but must be sufficiently strongly adsorbed by the textile fibres for an appreciable concentration to build up during washing and to remain after rinsing.

Over 200 chemically different optical brighteners are commercially available, and the choice for a given detergent depends on the type of textile and washing conditions for which the detergent is intended. Cellulosic fibres (e.g. cottons) possess adsorption characteristics different from those of synthetic fibres, and the brightener must be selected accordingly; in many cases, several brighteners may be added to the detergent to give a wide spectrum of activity.

Cellulosic fibres are hydrophilic and swell in water so that the pores in the amorphous region grow to around $15-30\,\text{Å}$ in diameter, large enough to admit the brightener molecules. High affinity for the fibre is obtained if the brightener possesses several conjugated double bonds and aromatic nuclei of coplanar configuration: fortunately, this is also just the requirement for a high fluorescence yield (cf. Section 4.3, pp. 74–75). Almost all brighteners now used for cellulosic fibres are derivatives of bis-triazinyl-4,4'-diamino-stilbene-2,2'-disulphonic acid (Fig. 8.20): the *trans* isomer is the one adsorbed. The various substituents have little effect on the emission spectra but alter the characteristics of adsorption onto the fibre.

Brightening of hydrophobic fibres (e.g. nylon, polyester, and acetate) takes place in a manner similar to the dyeing of these fibres, possibly involving the penetration of the molecules into canals between the fibre molecules, or,

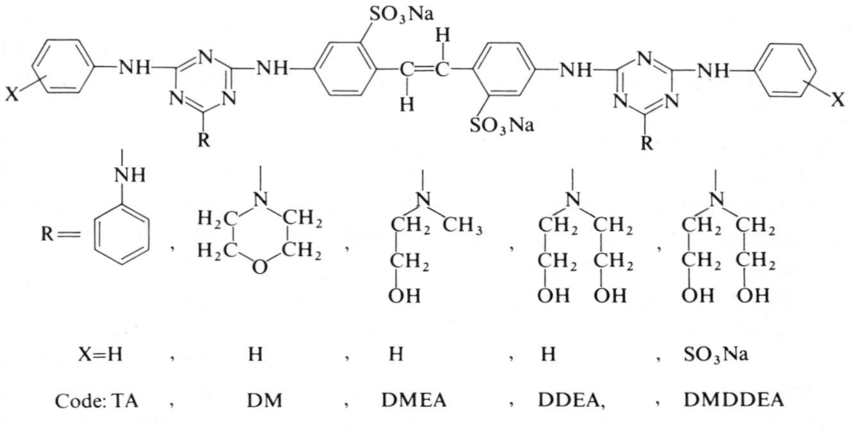

Fig. 8.20 Formulae of optical brighteners (bis-triazinyl-4,4'-diamino-stilbene-2,2'-disulphonic acid derivatives) suitable for cellulosic (e.g. cotton) fibres.

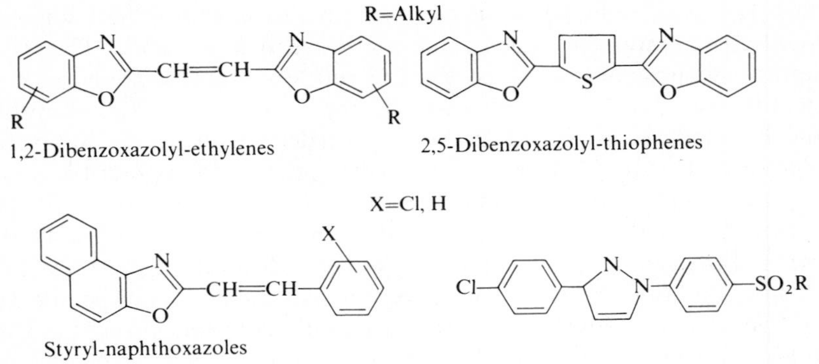

R=Alkyl

1,2-Dibenzoxazolyl-ethylenes

2,5-Dibenzoxazolyl-thiophenes

X=Cl, H

Styryl-naphthoxazoles

Fig. 8.21 Brighteners suitable for polyamide (e.g. Nylon), polyester (e.g. Terylene), and acrylic (e.g. Orlon) fibres.

alternatively, as a result of actual solution of brightener in the solid fibre. Figure 8.21 shows some brighteners active for polyamide, polyester, and acrylic fibres.

8.12 Photomedicine

We conclude our survey of the applications of photochemistry with a very brief reminder that photochemistry has important uses in medicine. Ultraviolet radiation is employed in disinfection, sterilization, and the purification of water. Fluorescence is used diagnostically in dermatological and dental practice. Ultraviolet curing of dental resins was mentioned in Section 8.8.2, and lightweight orthopaedic casts based on photopolymerization have also been reported. *Phototherapy* is concerned with the treatment of disease. Minor skin ailments often respond well to exposure to ultraviolet radiation. Psoriasis is a serious skin condition, and is treated by *photochemotherapy*: exposure to ultraviolet radiation is augmented by use of a photosensitizing drug, such as 8-methoxypsoralen, taken by the patient some hours before the exposure. Sometimes the ultimate effect of the ultraviolet radiation is experienced in different parts of the body from those exposed. The production of vitamin D_3 described in Section 8.10 as an industrial synthesis, proceeds in the body in the same way, 7-dehydrocholesterol in the skin being converted photochemically to the vitamin. Normally, exposure to sunlight produces sufficient vitamin D_3, which is essential for healthy bone formation. In patients forced to remain indoors, supplementary exposure to artificial ultraviolet radiation may provide protection against the development of fragile bones and rickets. Several other examples of photochemotherapy have been described, and this rapidly developing field affords an excellent example of the chemical effects of light being harnessed to the benefit of Man, and also provides a suitably optimistic note on which to finish this introduction to the study of photochemistry.

Bibliography

Roberts, R., Ouellette, R. P., Muradaz, M. M., Cozens, R. F., and Cheremisinoff, P. N. (1984). *Applications of photochemistry*. Technomic Publishing, Lancaster, PA.
Wald, G. (1959). Life and light. *Sci. Am.* **201**(4), 92.
Coyle, J. D. (1980). Light and biological systems. *Chem. Br.* **16**, 460.
Hendricks, S. B. (1968). How light interacts with living matter. *Sci. Am.* **219** (3), 174.

Special topics

Section 8.2.1: origin and evolution of the atmosphere
Wayne, R. P. (1985). Evolution and change in atmospheres and climates. Chapter 9 in *Chemistry of atmospheres*. Oxford University Press, Oxford.
Levine, J. S. (1985). The photochemistry of the early atmosphere. In J. S. Levine, ed., *The photochemistry of atmospheres* (ed. J. S. Levine) Chapter 1. Academic Press, Orlando, FL.
Wayne, R. P. (1988). Origin and evolution of the atmosphere. *Chem. Br.* **24**, 225.

Section 8.2.2: the stratosphere

Wayne, R. P. (1985). Ozone in Earth's stratosphere. Chapter 4 in *Chemistry of atmospheres*, Oxford University Press, Oxford.
Turco, R. P. (1985). The photochemistry of the stratosphere. In *The photochemistry of atmospheres* (ed. J. S. Levine) Chapter 3. Academic Press, Orlando, FL.
World Meteorological Organization (1981, 1986). *The stratosphere 1981* and *Atmospheric ozone 1985*. WMO, Geneva.
Anderson, J. G. (1987). Free radicals in the Earth's atmosphere. *Ann. Rev. Phys. Chem.* **38**, 489.
Stolarski, R. S. (1988). The Antarctic ozone hole. *Sci. Am.* **258**(1), 20.

Section 8.2.3: the troposphere

Wayne, R. P. (1985). The Earth's troposphere. Chapter 5 in *Chemistry of atmospheres* Oxford University Press, Oxford.
Graedel, T. E. (1985). The photochemistry of the troposphere. In *The photochemistry of atmospheres* (ed. J. S. Levine) Chapter 2. Academic Press, Orlando, FL.
Finlayson-Pitts, B. J. and Pitts, J. N., Jr (1986). *Atmospheric chemistry*. John Wiley, Chichester and New York.
Seinfeld, J. L. (1986). *Atmospheric chemistry and physics of air pollution*. John Wiley, Chichester and New York.

Section 8.3: photosynthesis
Sauer, K. (1979). Photosynthesis—the light reactions. *Ann. Rev. Phys. Chem.* **30**, 155.
Youvan, D. C. and Marrs, B. L. (1987). Molecular mechanisms of photosynthesis. *Sci. Am.* **256**(6), 42.

Foyer, C. H. (1986). The regulation of carbon assimilation in photosynthesis. *Chem. Br.* **22**, 723.

Miller, K. R. (1979). The photosynthetic membrane. *Sci. Am.* **241**(4), 102.

Calvin, M. (1962). The path of carbon in photosynthesis. *Angew. Chem.* **2**, 65.

Thomas, J. B. (1965). *Primary processes in biology*, Chapter 4. North-Holland, Amsterdam.

Porter, G. (1982). Photosynthesis. In *Light, chemical change and life* (eds J. D. Coyle, R. R. Hill, and D. R. Roberts) Chapter 6.3. The Open University Press, Milton Keynes.

Wolken, J. J. (1986). *Light and life processes*, Chapter 8. Van Nostrand Reinhold, New York.

Porter, G. (1986). Photosynthesis—the first nanosecond. *Pure Appl. Chem.* **58**, 1171.

Blankenship, R. E. (1981). Chemically induced magnetic polarization in photosynthetic systems. *Acc. Chem. Res.* **14**, 163.

Warden, J. T. and Bolton, J. R. (1974). Light-induced paramagnetism in photosynthetic systems. *Acc. Chem. Res.* **7**, 189.

Section 8.4: vision

Stryer, L. (1987). The molecules of visual excitation. *Sci. Am.* **257**(1), 32.

Schnapf, J. L. and Baylor, D. A. (1987). How photoreceptor cells respond to light. *Sci. Am.* **256**(4), 32.

Wolken, J. J. (1986). *Light and life processes*, Chapter 12. Van Nostrand Reinhold, New York.

Photobiochemistry and Photobiophysics, Vol. 13, special issue on vision (December) (1986).

Nathans, J., Thomas, D., and Hogness, D. S. (1986). Molecular genetics of human colour vision: the genes encoding blue, green, and red pigments. *Science* **232**, 193.

Masland, R. H. (1986). The functional structure of the retina. *Sci. Am.* **255**(6), 90.

Abrahamson, E. W. and Ostroy, S. E. (eds) (1981). *Molecular processes in vision* (Benchmark papers in biochemistry, Vol. 3). Hutchinson Ross, Stroudsberg, PA.

Dartnall, H. J. A. (ed.) (1972). *Photochemistry of vision*. Springer-Verlag, Berlin.

Section 8.5: photoimaging

Brinckman, E., Delzenne, G., Poot, A., and Willems, J. (1978). *Unconventional imaging systems*. Focal Press, London.

Phillips, R. (1983). *Sources and applications of ultraviolet radiation*. Academic Press, London.

Jacobson, R. E. (1980). Photochemical imaging systems. *Chem. Br.* **16**, 468.

Delzenne, G. A. (1979). Organic photochemical imaging systems. *Adv. Photochem.* **11**, 1.

Clark, M. G. (1985). Materials for optical storage. *Chem. Ind.* (15 April), 258.

Ledger, M. (1982). Non-conventional photoimaging. In *Light, chemical change and life* (eds J. D. Coyle, R. R. Hill, and D. R. Roberts), Chapter 5.12. The Open University Press, Milton Keynes.

Section 8.6: photography

James, T. H. (ed.) (1977). *The theory of the photographic process*, 4th edn. Collier Macmillan, London.
James, T. H. (1986). Chemical sensitization, spectral sensitization, and latent image formation in silver halide photography. *Adv. Photochem.* **13**, 329.

Section 8.7: photochromism

Dessauer, R. and Paris, J. P. (1963). Photochromism. *Adv. Photochem.* **1**, 275.
Brown, G. H. (ed.) (1971). *Photochromism. Techniques of chemistry*, Vol. III. Wiley–Interscience, New York.
Ledger, M. (1982). Non-conventional photoimaging. In *Light, chemical change and life* (eds J. D. Coyle, R. R. Hill, and D. R. Roberts), Chapter 5.12. The Open University Press, Milton Keynes.

Section 8.8.1 and 8.8.2: photochemistry of polymers (imaging and curing)

Guillet, J. (1985). *Polymer photophysics and photochemistry*. Cambridge University Press, Cambridge.
Phillips, R. (1983). *Sources and applications of ultraviolet radiation*. Academic Press, London.
Green, G. E. and Stark, B. P. (1981). Photopolymer systems and their applications. *Chem. Br.* **17**, 228.
Allen, N. S. and McKellar, J. F. (1980). Polymer–light interactions. *Chem. Br.* **16**, 480.
Phillips, R. (1984). Photopolymerization. In *Photochemistry past present and future* (eds R. P. Wayne and J. D. Coyle) p. 79. Elsevier Sequoia, Lausanne [also published as *J. Photochem.* **25**, Part 1 (1984)].
Roberts, E. D. (1985). Resists used in lithography. *Chem. Ind.* (15 April), 251.
Hageman, H. J. (1985). Photoinitiators for free radical polymerization. *Prog. Org. Coat.* **13**, 123.
Pape, M. (1975). Industrial applications of photochemistry. *Pure Appl. Chem.* **41**, 535.
Roffey, C. G. (1982). *Photopolymerization of surface coatings*. John Wiley, Chichester.
Curtis, H., Irving, E., and Johnson, B. F. G. (1986). Organometallic photoinitiated polymerisations. *Chem. Br.* **22**, 327.
Cundall, R. B. and Salim, M. S. (1982). Photoinitiation of polymerisation. In *Light, chemical change and life* (eds J. D. Coyle, R. R. Hill, and D. R. Roberts) Chapter 5.5. The Open University Press, Milton Keynes.
Ledwith, A. (1982). Photochemical cross-linking of polymers. In *Light, chemical change and life* (eds J. D. Coyle, R. R. Hill, and D. R. Roberts) Chapter 5.6. The Open University Press, Milton Keynes.
Jacobson, R. (1982). Photopolymers for imaging and resists. In *Light, chemical change and life* (eds J. D. Coyle, R. R. Hill, and D. R. Roberts) Chapter 5.13. The Open University Press, Milton Keynes.
Eaton. D. F. (1986). Dye sensitized photopolymerization. *Adv. Photochem.* **13**, 427.

Section 8.8.3: photodegradation and photostabilization

Scott, G. (1984). Photodegradation and photostabilization of polymers. In *Photochemistry past present and future* (eds R. P. Wayne and J. D. Coyle) p. 83. Elsevier Sequoia, Lausanne [also published as *J. Photochem.* **25**, Part 1 (1984)].

Grassie, N. and Scott, G. (1985). *Polymer degradation and stabilisation.* Cambridge University Press, Cambridge. TPL56. P6.67 (lem QD381.8.D48 Fore

Mellor, J. (1982). Chemistry of the photodegradation of polymers. In *Light, chemical change and life* (eds J. D. Coyle, R. R. Hill, and D. R. Roberts) Chapter 5.7. The Open University Press, Milton Keynes. QD714.L53 chem missing

Scott, G. (1985). Antioxidants *in vitro* and *in vivo*. *Chem. Br.* **21**, 648.

Section 8.9: solar energy storage

Hann, R. A. (1980). Solar energy conversion. *Chem. Br.* **16**, 474.

Wrighton, M. S. (1979). Photochemistry. *Chem. Eng. News* (3 Sept), 29.

Porter, G. (1982). Criteria for solar energy conversion. In *Light, chemical change and life* (eds J. D. Coyle, R. R. Hill, and D. R. Roberts), Chapter 6.1. The Open University Press, Milton Keynes.

Section 8.9.1: photochemical energy storage

Laird, T. (1982). Photochemical energy storage not based on redox systems. In *Light, chemical change and life* (eds J. D. Coyle, R. R. Hill, and D. R. Roberts) Chapter 6.2. The Open University Press, Milton Keynes.

Harriman, A. (1984). Artificial Photosynthesis. In *Photochemistry past present and future* (eds R. P. Wayne and J. D. Coyle) p. 33. Elsevier Sequoia, Lausanne [also published as *J. Photochem.* **25**, Part 1 (1984)].

Ryason, P. R. (1980). Solar photochemical fuel formation. *Surv. Prog. Chem.* **9**, 89.

Section 8.9.2: photoelectrochemical energy storage

Memming, R. (1988). Photochemical solar energy conversion. *Top. Curr. Chem.* **143**, 79.

Wrighton, M. S. (1979). Photochemical conversion of optical energy to electricity and fuels. *Acc. Chem. Res.* **12**, 304.

Bard, A. J. (1980). Photoelectrochemistry. *Science* **207**, 139.

Heller, A. (1981). Conversion of sunlight into electrical power and photoassisted electrolysis of water in photoelectrochemical cells. *Acc. Chem. Res.* **14**, 154.

Parkinson, B. (1984). On the efficiency and stability of photoelectrochemical devices. *Acc. Chem. Res.* **17**, 431.

Albery, W. J. (1982). Development of photogalvanic cells for solar energy conversion. *Acc. Chem. Res.* **15**, 142.

Peter, L. M. (1984). Electrochemical solar energy conversion. In *Photochemistry past present and future* (eds R. P. Wayne and J. D. Coyle) p. 37. Elsevier Sequoia, Lausanne [also published as *J. Photochem.* **25**, Part 1 (1984)].

Peter, L. M. (1982). Photoelectrochemical cells and their potential for the conversion and storage of solar energy. In *Light, chemical change and life* (eds J. D. Coyle,

R. R. Hill, and D. R. Roberts) Chapter 6.5. The Open University Press, Milton Keynes.

Hamakawa, Y. (1987). Photovoltaic power. *Sci. Am.* **256**(4), 76.

Section 8.10: photochemistry in synthesis

Phillips, R. (1983). *Sources and applications of ultraviolet radiation.* Academic Press, London.

Margaretha, P. (1982). Preparative organic photochemistry. *Top. Curr. Chem.* **103**, 1.

Clements, A. D. (1980). Photochemistry and commercial synthesis. *Chem. Br.* **16**, 464.

Carless, H. A. J. (1980). Photochemistry in laboratory synthesis. *Chem. Br.* **16**, 456.

Fischer, M. (1978). Industrial applications of photochemical syntheses. *Angew. Chem., Int. Ed. Engl.* **17**, 16.

Pape, M. (1975). Industrial applications of photochemistry. *Pure Appl. Chem.* **41**, 535.

Cocker, J. D. and Roberts, S. M. (1984). Relevance of photochemical transformations to the laboratory synthesis of biologically interesting molecules. In *Photochemistry past present and future* (eds R. P. Wayne and J. D. Coyle) p. 73. Elsevier Sequoia, Lausanne [also published as *J. Photochem.* **25**, Part 1 (1984)].

Pfoertner, K. H. (1984). Photochemistry in industrial synthesis. In *Photochemistry past present and future* (eds R. P. Wayne and J. D. Coyle) p. 91. Elsevier Sequoia, Lausanne [also published as *J. Photochem.* **25**, Part 1 (1984)].

Carless, H. (1982). Photochemistry in fine chemical manufacture. In *Light, chemical change and life* (eds J. D. Coyle, R. R. Hill, and D. R. Roberts) Chapter 5.9. The Open University Press, Milton Keynes.

Horspool, W. (1982). Photochemistry in large-scale synthesis. In *Light, chemical change and life* (eds J. D. Coyle, R. R. Hill, and D. R. Roberts) Chapter 5.10. The Open University Press, Milton Keynes.

Mellor, J. (1982). Photochemistry in laboratory synthesis. In *Light, chemical change and life* (eds J. D. Coyle, R. R. Hill, and D. R. Roberts) Chapter 5.11. The Open University Press, Milton Keynes.

Beddard, G. (1982). Photochemical isotope enrichment. In *Light, chemical change and life* (eds J. D. Coyle, R. R. Hill, and D. R. Roberts) Chapter 5.14. The Open University Press, Milton Keynes.

Section 8.11: optical brighteners

Zollinger, H. (1987). *Color chemistry.* VCH, Weinheim.

Stensby, P. S. (1968). Optical brighteners as detergent additives. *J. Am. Oil Chem. Soc.* **45**, 497.

Pape, M. (1975). Industrial applications of photochemistry. *Pure Appl. Chem.* **41**, 535.

Section 8.12: photomedicine

Phillips, R. (1983). *Sources and applications of ultraviolet radiation.* Academic Press, London.

Van den Bergh, H. (1986). Light and porphyrins in cancer therapy. *Chem. Br.* **22**, 430.

Appendix 1 Fundamental physical constants of importance in photochemistry

Symbol	Constant	Value
c	Speed of light *in vacuo*	$2.998 \times 10^8 \text{ ms}^{-1}$
h	Planck's constant	$6.63 \times 10^{-34} \text{ J s}$
k	Boltzmann constant	$1.38 \times 10^{-23} \text{ JK}^{-1}$
N	Avogadro's number	$6.022 \times 10^{23} \text{ mol}^{-1}$
R	Gas constant	$8.31 \text{ J mol}^{-1} \text{ K}^{-1}$
m	Mass of electron at rest	$9.11 \times 10^{-31} \text{ kg}$
e	Charge of electron	$1.60 \times 10^{-19} \text{ C}$

Appendix 2 Conversion table for energy units

A	B			
	J mol^{-1}	cal mol^{-1}	eV	cm^{-1}
J mol^{-1}	1	2.390×10^{-1}	1.036×10^{-5}	8.359×10^{-2}
cal mol^{-1}	4.184	1	4.336×10^{-5}	3.498×10^{-1}
eV	9.649×10^4	2.306×10^4	1	8.066×10^3
cm^{-1}	1.196×10	2.859	1.240×10^{-4}	1

Multiply value given in units under 'A' by figure in the same row of 'B' to get the value in the units of the relevant column of 'B'. For example, to convert $150\,000 \text{ J mol}^{-1}$ to eV multiply by 1.036×10^{-5} to give the result 1.554 eV.

INDEX

absorbed intensity, *see* intensity
absorption 17
  banded 34
  biphotonic *see* biphotonic absorption
  continuous 36, 39
  diffuse 39
  laser 168
  multipass, *see* multipass absorption
  resonance 168–9
  saturation 159
  singlet–triplet 83
  triplet–triplet 82, 168, 173
absorption coefficient 20–3, 41, 56, 138
absorption spectroscopy and spectrum 9, 33,
    81, 83, 168, 172, 177, 199, 230, 248; *see*
    also absorption
abstraction reactions 127, 142–6, 225
acceptor, *see* energy transfer, intermolecular
acetaldehyde (ethanal) 42–3
acid-base properties 127–9
actinometer, *see* chemical
    actinometer
activation energy 4, 72, 91–2, 97, 119, 130,
    137, 188, 213, 218, 230, 231
addition reactions 126, 127, 146–51, 153
adenosine triphosphate (ATP) 92, 198–200,
    209
adiabatic reactions 102–4, 107, 118, 129,
    135–7, 152; *see also* correlation of states
aerosols, atmospheric 196–7
afterglow 67; *see also* air afterglow
air afterglow 96, 167
air pollution 189–91, 193–7, 251
airglow 98
aldehydes 53, 148, 192, 196, 209–10; *see also*
    carbonyl compounds
alignment of photofragments 176, 180
alkenes 51, 146–7, 148, 149–50
allowed transition 22; *see also* selection rule
ammonia 231, 243
amphibians 184
amplification of radiation 120
  in visual process 209
angular momentum 24, 27, 103, 132, 170
animal life 184–5
anionic polymerization, *see* ionic
    polymerization
anode, *see* electrodes
antarafacial shifts 141
Antarctic ozone hole 191, 251

anthracene 77, 80, 112, 114–16, 117
anti-Stokes lines 76, 116–17, 173
argon ion laser 122, 157, 158
aromatic transition state 133
atmosphere
  origin and evolution of 182–5, 251
  pollution of 188–91, 193–7, 251
atmospheric photochemistry 181–97, 251
atom recombination reactions, *see*
    recombination reactions
automobile exhaust gases, photochemical
    pollution by 193–7
autoxidation 227–9
auxiliary pigments
  in photosynthesis 199; *see also* carotene;
    chlorophyll; pigment systems in
    photosynthesis
azo dye, *see* dyes

bacteria 184
balloon measurements 189, 251
band gap 236–42
bathorhodopsin, *see* pigments in vision
Beer–Lambert law 20–3, 161
  breakdown of 60
benzene 27, 28, 51–2, 80, 107, 140, 147, 245
benzophenone 104–5, 108–9, 119, 127–8,
    131, 143–5, 225, 228, 235
Becquerel phosphorimeter, *see*
    phosphorimeter, Becquerel
biacetyl 104, 107, 119
biphotonic absorption 59, 117; *see also*
    multiphoton processes
Birge–Sponer extrapolation 35–6
black-body radiation 2, 157, 159
blue shift 44–5
Bohr condition 18
Boltzmann distribution law 4
bond energy 4
Born–Oppenheimer approximation 28, 30
brain stimulation in visual process 209, 252
breaking off of emission bands 39
bridge reactions in photosynthesis 203–4,
    251–2; *see also* electron relay
bromine, molecular 41
buta-1,3-diene 133–7, 146–7
butanal 43

7 D Rec.